SEMICONDUCTOR MEMORIES

CONTRIBUTORS

ANDREW R. BERDING, *Advanced Memory Systems, Inc., Sunnyvale, California*

COLIN CLIFFORD, *General Instrument Microelectronics Limited, Fife, Scotland*

H. G. DILL, *Hughes Aircraft Company, Newport Beach, California*

DONALD V. DiMASSIMO, *G-V Controls, Inc., Livingston, New Jersey*

JERRY EIMBINDER, *EEE Magazine, New York, New York*

MARCIAN E. HOFF, JR., *Intel Corporation, Mountain View, California*

ANTHONY HOLBROOK, *Computer Microtechnology, Inc., Sunnyvale, California*

MERLE V. HOOVER, *RCA Corporation, Somerville, New Jersey*

ROBERT JUAN, *Signetics Corporation, Sunnyvale, California*

ARTURO KRUEGER, *Motorola Semiconductor Products Division, Geneva, Switzerland*

E. FLOYD KVAMME, *National Semiconductor Corporation, Santa Clara, California*

DAVID N. LEONARD, *Texas Instruments Incorporated, Dallas, Texas*

ROBERT C. LUTZ, *Advanced Memory Systems, Inc., Sunnyvale, California*

RICHARD B. MANN, *Texas Instruments Incorporated, Dallas, Texas*

MICHAEL R. McCOY, *Electronic Arrays, Inc., Mountain View, California*

ANH NGUEYEN-HUU, *Transitron Electronic Limited, Maidenhead, Great Britain*

ROBERT NORMAN, *Nortec Electronics Corporation, Santa Clara,* California

CARROLL R. PERKINS, *Hughes Aircraft Company, Newport Beach,* California

MELVIN G. SNYDER, *Raytheon Company, Mountain View, California*

GEORGE STEVENSON, *General Instrument Microelectronics Limited, Fife, Scotland*

EDWARD F. TARBOX, *Sylvania Electric Products, Inc., Woburn, Massachusetts*

T. N. TOOMBS, *Hughes Aircraft Company, Newport Beach, California*

SEMICONDUCTOR MEMORIES

JERRY EIMBINDER, *Editor*

Wiley-Interscience

Division of John Wiley & Sons, Inc.

New York · London · Sydney · Toronto

Library of Congress Catalog Card Number: 72-148501

ISBN 0 471 23460 5

Printed in the United States of America

10 9 8 7 6 5 4 3 2 1

Preface

In this book every type of semiconductor memory in use or in active development is described, and actual or potential applications are explored.

The material illustrates that the question of which semiconductor-memory technologies are best has given rise to widely differing opinions in the semiconductor-device industry. For example, the reader will find that Marcian E. Hoff favors silicon-gate structures, Carroll R. Perkins is an ion-implantation advocate, and Merle V. Hoover prefers the COS MOS (complementary-symmetry metal oxide semiconductor) devices. Indeed, the various MOS and bipolar technologies covered here represent many views.

A considerable amount of knowledge about semiconductor memories has been gathered in this volume to help the system designer choose the optimum semiconductor-memory approach for his application.

Because the semiconductor-memory explosion is making decision making much more complicated for the IC user, it is hoped that this book will help him to better deal with the rapidly changing semiconductor-memory field. A quick survey of the current activity in the semiconductor-memory IC field is presented in the first chapter.

Marcian E. Hoff compares various semiconductor memories in Chapter 2 but concentrates on the read/write memory. Hoff spends some time weighing MOS static and dynamic shift-register features, pointing out that the dynamic circuit is cheaper and faster but requires more elaborate drive circuitry. He moves on to the discussion of dynamic read/write random-access memory and explains the RAM's modes of organization and operation. He notes one RAM drawback—on a chip of fixed size, a read-only memory (ROM) permits greater bit capacity than does a RAM because of the small size of the ROM cell. To illustrate the interconnection of basic RAM packages, Hoff gives two construction examples, one for a 256-word by 32-bit/word memory, the other for a 4096-word by 12-bit/word memory.

The many ways in which a designer can add semiconductor memory to his system are covered by Robert Norman in Chapter 3. Norman comments on the decision to build or buy integrated circuits, on designing with standard and custom memory ICs, and on buying memory modules or subsystems.

When speed is essential, current-mode-logic (CML) is a serious contender for system circuitry. The advantages of CML are explored in Chapter 4 by Andrew R. Berding and Robert C. Lutz.

Ion-implanted techniques for building MOS memories are explained by Carroll R. Perkins, H. G. Dill, and T. N. Toombs in Chapter 5. Of particular interest is a comparison of the ion-implanted, standard MOS-*p*-channel and silicon-gate (self-registering gate) structures.

The MOS read-only memory is investigated in Chapter 6 by Michael R. McCoy, with emphasis placed on custom-bit-pattern ROMs. McCoy outlines a procedure for a customer-vendor relationship for simplified specifying of custom-bit-pattern ROMs.

In Chapter 7, David N. Leonard looks at MOS RAMs and MOS ROMs and discusses the application of both. The same author introduces the MOS content-addressable memory in Chapter 8 and presents an example illustrating how it functions.

In-depth treatment of complementary-symmetry MOS ICs is given in Chapters 9 and 10 by Donald V. DiMassimo and Merle V. Hoover, respectively. Because of the lack of extensive coverage of this area in literature, both chapters go beyond the relationship of COS MOS to memory and delve briefly into logic as well.

Chapter 11, written by George Stevenson, shows how MOS ICs can be used in various parts of a typical system. Colin Clifford looks at the interfacing of silicon-nitride devices with diode–transistor logic (DTL) and transistor–transistor logic (TTL) in Chapter 12.

A thorough discussion of character generation using static shift registers and read-only memories is made by Richard B. Mann in Chapter 13. In Chapter 14, E. Floyd Kvamme also applies static shift registers and read-only memories in the design of systems using microprogramming techniques.

A 8192-bit memory module that combines MOS and emitter-coupled logic is the subject of Chapter 15. Arturo Krueger explains how this device functions.

Melvin G. Snyder in Chapter 16 and Anh-Nguyen-huu in Chapter 17 introduce the basic bipolar semiconductor memory and describe its mode of operation and its use in building larger arrays.

Finally, Chapter 18 consists of a helpful rundown of definitions given by Robert Juan; and in Chapters 19 and 20, respectively, Edward F. Tarbox shows how to convert from binary to binary-coded decimal numbers using bipolar read-only memories and adders, and the case for using beam-lead devices in memory systems is presented by Anthony Holbrook.

For the most part, the chapters are based on papers given at seminars conducted by *EEE Magazine*. Most of the authors are applications engineers employed by integrated-circuit manufacturers.

Jerry Eimbinder

New York, New York
October, 1970

Contents

Chapter 17

Chapter 18

Chapter 19

SEMICONDUCTOR MEMORIES

1

Semiconductor-Memory Integrated Circuits

by Jerry Eimbinder

EEE Magazine

1.1 BASIC ELEMENTS

The basic element in bipolar and static MOS (metal oxide semiconductor) memories is the flip-flop. Employment of flip-flops as storage units makes it possible to combine a small amount of memory with drive, sense, and decode circuitry on the same chip. However, for larger memories, the demand on real estate is compounded by the additional requirement for peripheral decode and drive circuitry.

In a dynamic MOS integrated circuit (IC), the basic storage cell is a capacitor; its job is to hold or release stored charge. Because capacitors have leakage, the stored charge must be continually refreshed. A refreshing system, therefore, must be included in dynamic MOS circuitry.

A number of semiconductor devices have been developed for use as storage units in semiconductor memories. Besides the conventional bipolar transistor (see Figure 1.1) and the metal oxide semiconductor field-effect transistor (MOS FET), these elements include the silicon-gate, silicon-nitride, Schottky-modified, ion-implanted, metal-alumina-silicon, silicon-on-sapphire, and silicon-on-spinel transistors, and the charge-coupled-device.

Integrated-circuit designers have come up with many different variations

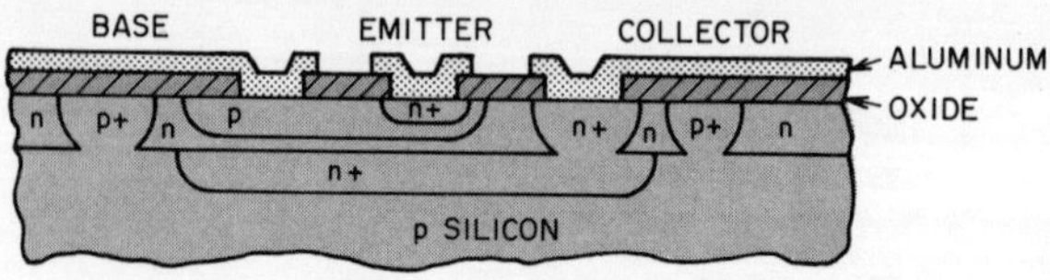

Figure 1.1. Bipolar transistor structure.

of concepts for producing semiconductor-memory ICs. As an illustration, ion implantation is used by one manufacturer to produce MOS FET channels. Another manufacturer ion implants parts of the source and drain. The combination of ion implantation with complementary-symmetry MOS has also been successfully accomplished.

Techniques for building semiconductor-memory ICs are in a continuous state of change. For example, silicon-gate MOS ICs have been developed with silicon nitride between the semiconductor material and the dioxide layer situated below the silicon gate, with silicon nitride on top of the silicon gate as a passivating material, and with no silicon nitride at all. The last of these three approaches is currently the popular one.

1.2 SILICON-GATE MOS INTEGRATED CIRCUITS

The key physical difference between a silicon-gate IC and a conventional MOS IC (see Figures 1.2 and 1.3) is the use of polycrystalline silicon instead of aluminum for the gate electrode of the former.

The work function of the silicon is lower than the work function of the aluminum commonly used in MOS ICs, and this results in substantially reduced threshold voltage. The silicon-gate MOS ICs that have become available since the introduction of the first silicon-gate standard IC in 1969 have used 1-1-1 crystal orientation, but other orientations have been investigated.

Fabrication begins with placement of an oxide (silicon dioxide) layer on the *n*-type silicon wafer. After the silicon dioxide layer is grown, windows are etched for the MOS transistors. Then the layer of polycrystalline silicone is deposited over the wafer.

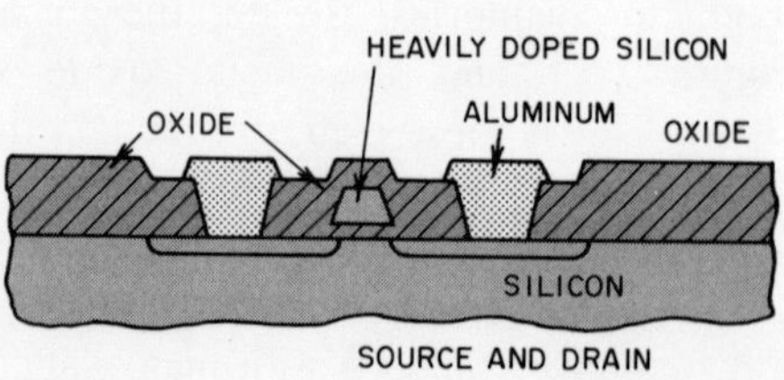

Figure 1.2. The silicon-gate MOS transistor.

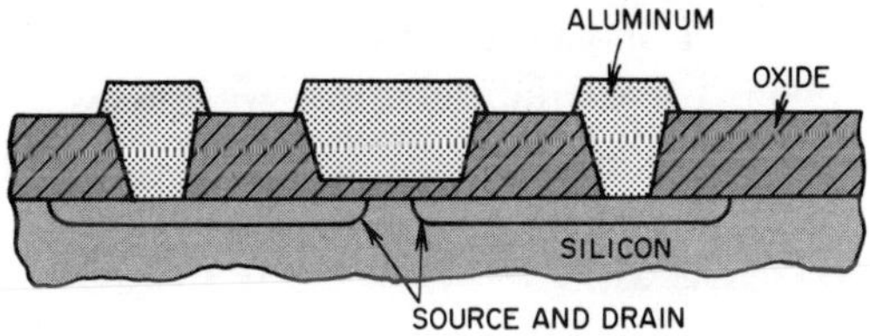

Figure 1.3. Conventional MOS transistor.

Next the polycrystalline silicon layer is etched to define the gate structure and open windows above the source-and-drain locations. This is followed by diffusion of a *p*-type dopant (such as boron) in the source-and-drain regions and formation of the source-and-drain junctions. The dopant also diffuses into the polycrystalline silicon gate, turning it into heavily *p*-doped silicon.

Another silicon dioxide layer is put down, and windows are opened for the source-and-drain contacts. Now the IC is ready for deposition of aluminum metalization over the entire surface.

The polycrystalline silicon that remains in the gate structure after etching sits on top of a dioxide region created during the first silicon dioxide deposition. The polycrystalline silicon thus protects this dioxide region from subsequent etching. This type of gate structure is called a self-aligning gate.

The conventional type of MOS IC gate structure must include allowance for misalignment of the three masks used for source-and-drain diffusion, gate dioxide growth, and gate-metal deposition. The conventional device, therefore, requires a considerably larger gate and more gate-metal deposition than are needed for the silicon-gate transistor.

To ensure that adequate gate metalization has been deposited, the gate aluminum layer of the conventional MOS transistor is permitted to overlap nearby portions of the diffused source-and-drain regions. This causes overlap capacitances and reduces switching-speed capability.

The combination of low threshold voltage with good speed capability thrust considerable attention upon the silicon-gate MOS IC in 1969. The low threshold voltage was particularly appealing because it made compatible operation with bipolar ICs possible.

The silicon-gate process is attractive to IC manufacturers because it promises high yields. The gate oxide is safeguarded by the layers of silicon and oxide above it; therefore greater freedom exists, should application of high temperature during subsequent processing be necessary. (In fabricating conventional MOS IC transistors, the gate oxide is exposed until the final metalization step.) This protection is important if bipolar transistors are being produced on the wafer, since their fabrication requires several high-temperature steps.

In addition, more silicon-gate transistors than conventional MOS transistors can be obtained from the same size wafer because of the smaller gate-area requirement of the former.

1.3 SILICON-NITRIDE MOS INTEGRATED CIRCUITS

The low threshold voltage necessary for compatibility with bipolar circuits can be achieved if a layer of silicon nitride is grown above the channel of the MOS transistor (see Figure 1.4).

The nitride, because its dielectric constant is higher than that of silicon dioxide (typically 7.5 compared with 3.9), increases the capacitance between the channel and the gate. Threshold voltage is dependent on a number of variables, including this capacitance. Increasing this capacitance lowers threshold voltage while maintaining, from a mechanical standpoint, a practical insulation width between gate and channel.

Silicon nitride can be used in combination with silicon dioxide as the gate dielectric. This makes processing more difficult, but it combats the hysteresis effect that may occur when silicon nitride is used alone. It also reduces the possibility of degradation from pinholes in either nitride or dioxide layers, since the possibility that two pinholes will line up is quite remote.

The use of silicon nitride also makes possible the use of 1-1-1 oriented silicon crystal (the orientation normally employed for MOS transistors). The use of 1-0-0 orientation can be resorted to in order to reduce threshold voltages, but parasitics are a more serious problem when it is used. Other methods of reducing threshold voltage include the previously mentioned use of a silicon gate and the use of an ion-implanted channel.

1.4 COMPLEMENTARY-SYMMETRY MOS INTEGRATED CIRCUITS

In conventional bipolar and *p*-channel MOS ICs, most of the power dissipation is quiescent power contributed by the load resistor. In the complementary-symmetry circuit (see Figure 1.5), the transistor's load resistor is replaced by a transistor of opposite polarity. This results in a considerably lower level of dissipation of quiescent power.

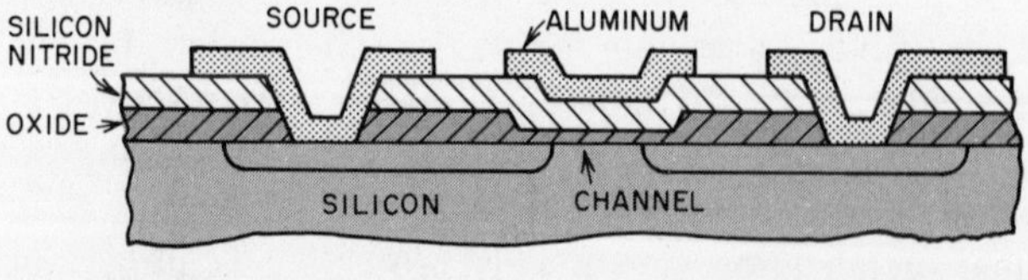

Figure 1.4. MOS transistor with silicon-nitride layer.

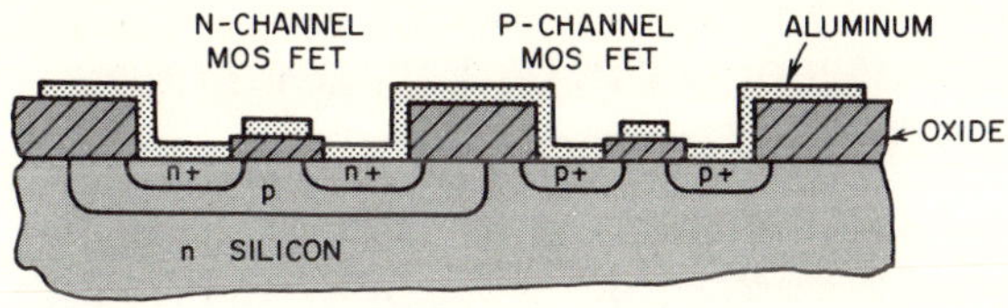

Figure 1.5. Complementary-symmetry (*n*-channel, *p*-channel) MOS construction.

An *n*-type silicon substrate is used in the fabrication of complementary-symmetry MOS transistors. A *p*-type diffusion creates a substrate for the *n*-channel transistors; the *n*-type wafer acts as the substrate for the *p*-channel devices. Both the *p*-channel and the *n*-channel transistors are enhancement (normally off) MOS devices.

Good operating speeds are expected from devices built with the complementary-symmetry process because the output node capacitance is always charged and discharged through the "on" transistor.

Two devices (an *n*-channel and a *p*-channel) form a flip-flop; since one of the transistors is always conducting, a low impedance path always exists from the output to either the power supply or ground. This low impedance path minimizes capacitive noise pickup. The nonconducting transistor contributes to lowering of the standby power dissipation by blocking the current flow from power supply to ground.

1.5 COMPLEMENTARY-SYMMETRY MOS–BIPOLAR INTEGRATED CIRCUITS

By placing vertical and lateral *npn* bipolar transistors on the same chip with MOS FETs (see Figure 1.6), the capability to drive higher capacitance loads can be achieved. In addition, because bipolar transistors are used at the output, interfacing to bipolar circuits is not a problem.

The vertical and lateral transistors are connected to form a bipolar transistor with an isolated collector. An n^+ substrate and an *n*-type epitaxial layer are used below the *p*-type base regions of the transistors.

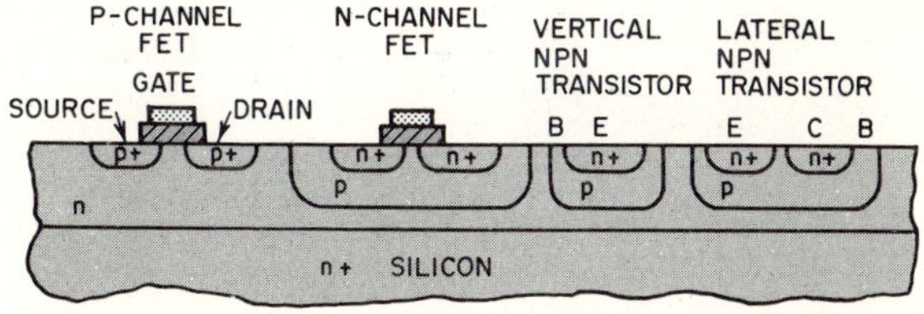

Figure 1.6. Complementary-Symmetry MOS–bipolar transistor arrangement.

1.6 SCHOTTKY-PROCESS INTEGRATED CIRCUITS

The announcement of commercially available Schottky-device ICs in 1969 created quite a stir because of the high speeds this technology appeared to offer. Researchers developing ICs incorporating Schottky devices reported that diode–transistor logic (DTL) ICs could achieve the same speeds provided by conventional transistor–transistor logic (TTL) and emitter-coupled logic (ECL).

In a typical Schottky-process IC, the Schottky diode is formed by the junction of the aluminum metalization and the *n*-type silicon below. This results in a lower forward voltage drop than that exhibited by discrete, diffused *pn* junction Schottky diodes.

Forward current, carried by electrons, flows out of the semiconductor material and into the aluminum. Because these electrons are brought into equilibrium with the electrons present in the metal, no stored charge is accumulated. By eliminating the buildup of stored charge, the device offers high-speed capabilities.

This absence of stored charge can be used to advantage if the Schottky diode is placed in parallel with the base-to-emitter junction of an *npn* transistor (the metal electrode is connected to the base and the *n* region of the collector). The diode clamps the transistor and takes most of the excess base current, thus preventing the transistor from saturating. No charge storage is built up in either the transistor or the diode.

Physically, the Schottky-diode transistor can be formed by letting the opening for the base contact of the transistor extend beyond the base region and over the collector region. Thus when the aluminum metalization fills the opening, it becomes both the base contact and one electrode of the Schottky diode. High field concentrations at the edges of the Schottky device can be avoided (1) by depositing the metal so that it extends beyond the opening and overlaps a sufficient amount to the dioxide layer as shown in Figure 1.7 or (2) by employing diffused guard rings below the edges of metalization as shown in Figure 1.8.

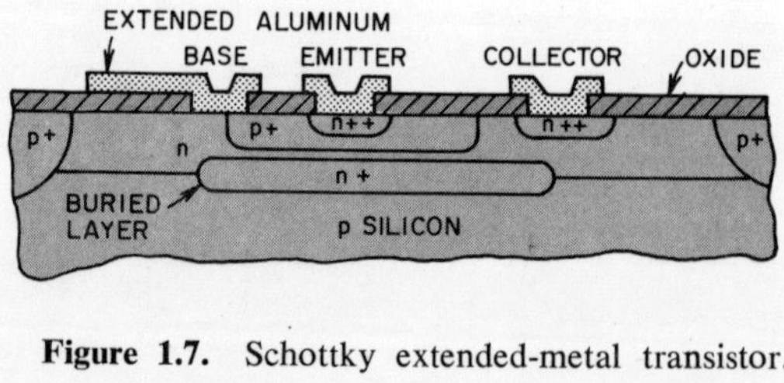

Figure 1.7. Schottky extended-metal transistor.

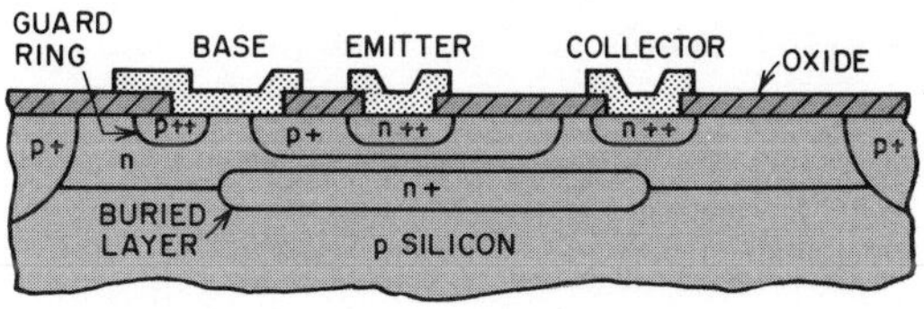

Figure 1.8. Schottky guard-ring transistor.

1.7 ION-IMPLANTED MOS INTEGRATED CIRCUITS

Ion implantation is a process in which dopant ions are accelerated to high energies and directed against a silicon wafer. How deeply the bombarding ions penetrate is determined largely by their energy content.

Ion implantation is used in various areas of a transistor, since more than one implanting technique has been derived. One approach calls for parts of the source and drain to be ion implanted (see Figure 1.9); another applies ion implantation only in the channel region (see Figure 1.10). In both approaches the implantation is carried out for areas covered with thin oxide. Areas having metalization, thick oxide, or photoresist reject implantation. Ion implantation in the source-and-drain regions can be accomplished in conjunction with a self-aligning gate and can be applied after the gate metalization has been put down.

The ion-implanted self-aligning-gate transistor has lower overlap capacitance than a self-aligning-gate transistor because the latter device experi-

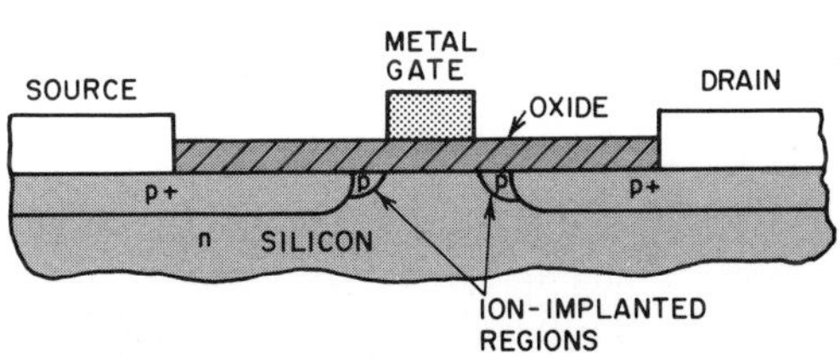

Figure 1.9. Ion-implanted MOS transistor.

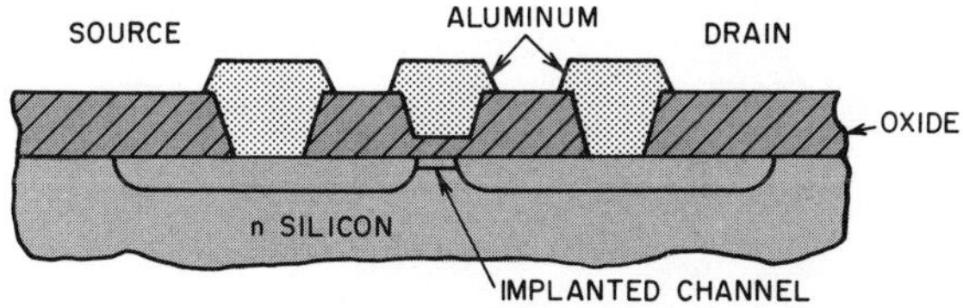

Figure 1.10. Implanted-channel MOS transistor.

ences some sideways diffusion (although not as much as a conventional MOS transistor). This makes the ion-implanted device a potentially faster device than either the conventional MOS transistor or a non-ion-implanted self-aligning-gate MOS transistor.

The implanting of ions in an MOS transistor's channel results in reduced charge per unit area and reduced threshold voltage. This technique can lower the threshold voltage enough so that the MOS transistor can be compatible with bipolar circuits.

1.8 METAL-ALUMINA-SILICON INTEGRATED CIRCUITS

A metal-alumina-silicon FET (see Figure 1.11) is the same as a conventional MOS IC FET except that alumina is used in place of oxide as the gate dielectric material.

The characteristics exhibited by the alumina can be varied by changing the voltage applied to the gate. The reaching of a critical applied voltage results in alteration of the device's threshold voltage. In a memory application, the shift from one threshold voltage to another can correspond to the storage of a bit of information.

Depending on the polarity of the applied voltage, electrons can be injected into the alumina either from the gate metal or from the silicon. Once the electrons are present in the alumina, a certain value of gate voltage must be reached in order for a channel to be created between the source and drain.

The change in threshold voltage is nonvolatile. Even if both the gate voltage and the power are removed, the threshold voltage remains a fixed characteristic.

Volatility refers to the loss of contents by a semiconductor memory when power is lost. One solution to this problem when volatile memories are employed is the use of battery backup. When a power failure occurs, the battery provides power for transfer of the contents to a nonvolatile memory such as a disk.

The "permanent" storage of threshold voltage level by the metal-alumina-silicon IC is actually erasable by application of radiation.

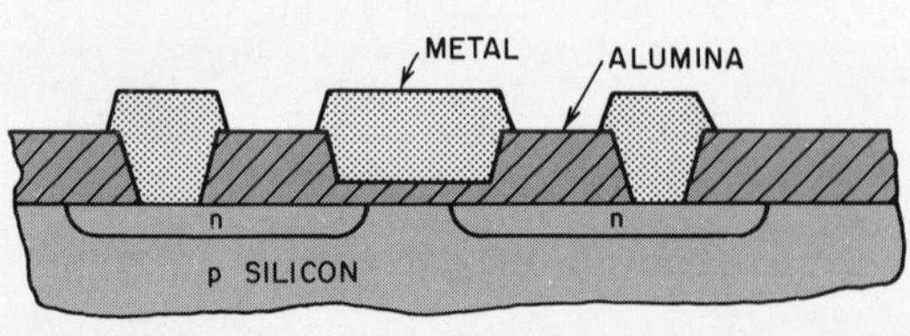

Figure 1.11. Metal-alumina-silicon transistor.

1.9 SILICON-ON-SAPPHIRE MOS INTEGRATED CIRCUITS

The use of sapphire rather than silicon as the underlying structure of an integrated circuit (see Figure 1.12) offers one very important advantage: extremely high insulation resistance between components of the IC. The decrease in substrate parasitic capacitance is so drastic that researchers claim that MOS circuits, if built on sapphire, could be as fast or faster than bipolar ICs.

Parasitic interaction between active elements is also virtually eliminated by silicon-on-sapphire (SOS) construction because of the superior isolation of the elements, which float like islands on a sea of sapphire.

Fabrication techniques under study for SOS ICs vary considerably. The main idea, however, is to put individual islands of silicon elements on the sapphire and connect the elements by metalization, thus forming a circuit.

Spinel, a crystalline material in many ways similar to sapphire but more compatible physically with silicon, is being considered as a possible substitute for silicon. Spinel, like silicon, has cubic crystal structure, whereas sapphire has rhombohedral crystalline form. Sapphire must undergo more critical surface preparation in order to induce single-crystal epitaxial silicon layers to grow on it. However, both types of substrates, because of their insulating properties, eliminate the requirement for IC transistors to have guard bands to ensure adequate *p–n* junction isolation from transistor to transistor.

Complementary MOS SOS circuits can be built employing the same number of diffusion steps needed to fabricate *p*-channel MOS ICs. This is possible because the dopants for both *n*-channel and *p*-channel devices can be applied simultaneously. A typical complementary MOS SOS fabrication process requires one more oxide deposition and one more etching step than the conventional *p*-channel MOS process used for silicon MOS ICs.

1.10 CHARGE-COUPLED-DEVICE MOS INTEGRATED CIRCUITS

The charge-coupled-device (CCD) MOS IC uses MOS capacitors to process injected charge (see Figure 1.13). There are no transistors on the

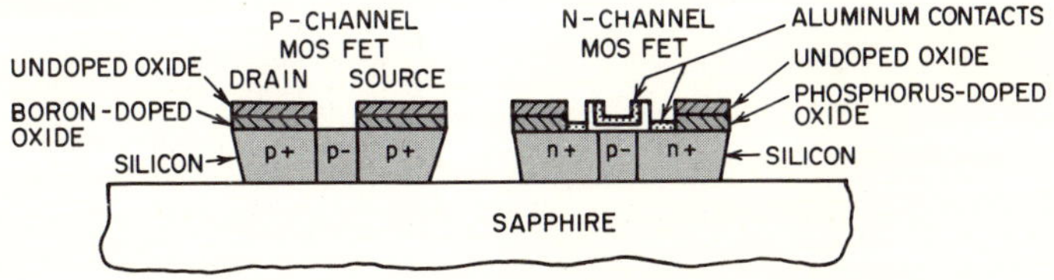

Figure 1.12. Silicon-on-sapphire transistor.

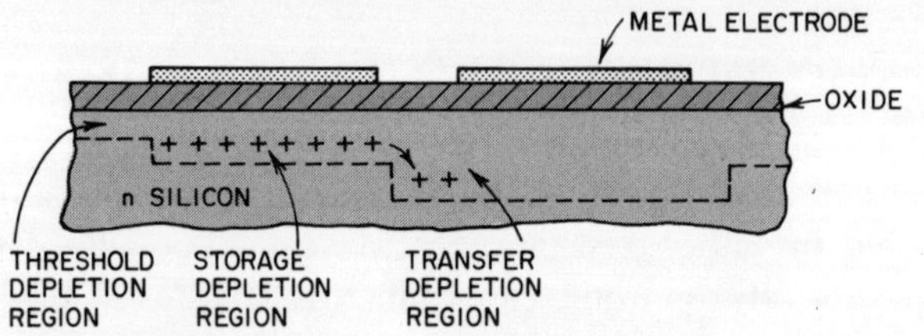

Figure 1.13. Charge-coupled device.

silicon substrate, no windows to be cut for diffusing dopant, and no *p–n* junctions in the structure.

Each charge-coupled device can produce minority carriers and store them in a special storage region below the interface of the semiconductor and the oxide layer. After they have been stored, the charges can be moved along the surface of the semiconductor to other locations, and then their presence or absence can be detected.

A biasing voltage is applied to the metal electrode of the MOS capacitor creating a depletion zone at the semiconductor–oxide interface. The application of a "storage" voltage greater than this bias voltage opens a well at this position and makes possible acceptance of carriers.

If minority carriers flow into the silicon, they collect at the semiconductor surface in the well. This accumulation of charge may be moved to an adjacent capacitor by placing a larger "transfer" voltage on the metal electrode.

The larger voltage deepens the well, and the carriers are attributed to the deeper well and transfer to the new electrode.

Once the charge is transferred, the voltage on the first MOS capacitor is returned to the normal biasing potential and the voltage on the receiving capacitor is lowered to the storage potential. The process can now be repeated by transferring the charge stored by the second capacitor to another adjacent capacitor.

2

Application Considerations for Read/Write Memories

by Dr. Marcian E. Hoff, Jr.

Intel Corporation

Memories built with semiconductor devices are becoming competitive from a cost standpoint with ferrite-core memories. In addition, they offer superior performance and greater freedom of organization.

The economic considerations associated with semiconductor-memory fabrication are quite different from those associated with magnetic memories. The optimum semiconductor-memory module size is determined by the number of chips yielded per wafer, the cost of processing the wafer, the cost of the package, and the losses associated with packaging. Unlike with magnetic memories, it is possible with semiconductor memories to include the address decoding function as well as the memory function on the chip. Including the decoder on the chip reduces the number of lead bonds that must be made to the chip.

The inclusion of address decoders requires little additional space on the chip since bonding pad area is traded for decoder area. Theoretically, there is a loss in yield because additional active area is needed. However this is partially offset by improved assembly yields and reduced package costs associated with the use of packages of fewer leads. These factors now tend to make modules from 64 to 1024 bits the most economically attractive. As manufacturing techniques improve, larger modules may become more desirable. With small, decoded modules as building blocks, multiphase clocking and other parallel techniques may be used with relatively little increase in cost over more conventional modes of operation.

Semiconductor memories of many types, based on both MOS and bipolar technologies, are available. These memories differ in cost, organization, and performance. In general, MOS technology offers lower cost, slower performance memories than bipolar. The use of dynamic MOS circuits permits retention of data at very low power. Thus, when volatility of memory data is a problem, dynamic MOS devices may be used together with batteries to provide backup power.

Before studying the key considerations for applying semiconductor memories, it will be helpful to examine some of the characteristics of semiconductor-memory devices.

2.1 BIPOLAR MEMORIES

Most bipolar memories use a relatively standard flip-flop as the storage element. Variations in process techniques include the use of an epitaxial layer for resistors and the use of Schottky barrier diodes to improve response speed and for cell selection. Figure 2.1 shows two basic bipolar memory cells; one uses multiple-emitter transistors, the other uses Schottky-diode selection.

Bipolar memories currently available include 64-bit (16 words by 4 bits/word) devices, fully decoded, and 256-bit (256 by 1 bit) partially decoded memories. The 64-bit memory typically offers 60-ns access time and, if built with the cell of Figure 2.1*a*, has a typical power dissipation of 6 mW per bit. The 256-bit memory offers approximately 80-ns access time with power dissipation of less than 2 mW per bit if built with cells similar to the one shown in Figure 2.1*b*. Both memories can be contained in 16-lead packages. A commercially available 256-bit memory uses a

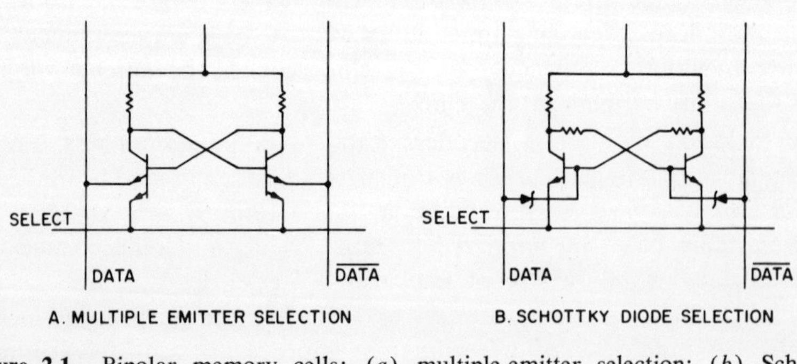

Figure 2.1. Bipolar memory cells: (*a*) multiple-emitter selection; (*b*) Schottky-diode selection.

3-out-of-6 code for both *X* and *Y* addresses, employing in each case only 16 out of the 20 possible codes. A decoder package for this code is available. Two decoder packages are sufficient to drive 32 memory packages organized as a 256-word, 32 bit/word memory.

Both the memories described offer "chip select" capability, and have data output leads that can be OR-tied. (The 256-bit memory uses a common input–output data lead.) Thus, additional address leads may be decoded to provide chip selection, permitting modules with larger numbers of words to be constructed easily, with a minimum number of additional parts.

With these components, small- to medium-sized memories with cycle times of 80 to 150 ns may be constructed. These memories are used most commonly as computer scratch-pads and as (Caché-type) buffer memories.

2.2 MOS MEMORIES

Many different MOS memories are available or in development. Both static and dynamic storage cells may be built using MOS technology, and both types of MOS storage may be organized as random-access memory (RAM) or as serial (shift-register) memory. Dynamic MOS shift-register memories usually offer the lowest cost per bit. Very little chip area is wasted, for the shift register can use small memory cells, requires no decoder, and needs very few leads. Static MOS memories usually exhibit poorer performance and cost more than dynamic ones, but they can be easier to drive since, generally, dynamic memories require clock signals in addition to power supplies.

Dynamic MOS circuits make use of the very low leakage current which is associated with the gate circuits and junctions of properly designed MOS devices. These leakage currents are small enough to permit the circuit parasitic capacitances to exhibit time constants from milliseconds to seconds.

A dynamic MOS inverter is compared with a static MOS inverter in Figure 2.2.

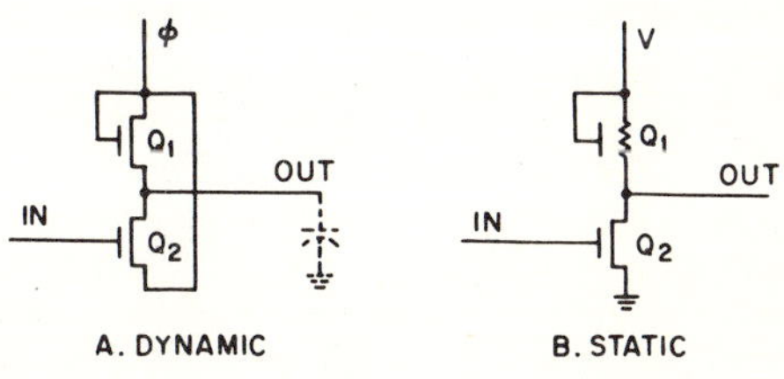

Figure 2.2. MOS inverters: (*a*) dynamic, (*b*) static.

In the static inverter, an MOS transistor is used for Q_1, a resistorlike device. When the input signal is a "1" (negative voltage) the output must be a "0." Thus the resistance of Q_1 must be very high when compared to that of the lower transistor, Q_2, or the output voltage will be too great to be considered a "0."

As this resistor device is turned on harder than the lower device, certain geometric relations between the two devices must be maintained. The resistor must have a long narrow channel, and the lower device must have a short wide channel. Since the minimum channel dimension tends to be determined by the minimum tolerance possible, both devices are larger than a minimum-size device (having an approximately square channel). In addition, much less current is available for charging an output load capacitance than for discharging it. Part of the load capacitance is a parasitic contribution from the inverter devices. Because these devices are larger than the minimum device, their load capacitances are greater, reducing the speed of the circuit.

In the dynamic MOS circuit, when an input Φ goes negative (for *p*-channel devices) the output is charged, via Q_1, to a negative potential. The only loads permitted are capacitive (except for junction leakage), so when Φ returns to ground potential, the output will remain negative unless Q_2 conducts. Device Q_2 will conduct only if it receives a "1" input. Therefore, when a pulse is applied on line Φ, the output after a short delay will represent the logical inverse of the input. Because circuit operation does not depend upon the ratio of device resistances, both devices can be of minimum geometry. Thus, some of the tradeoffs between static and dynamic circuits become obvious: the dynamic circuit is less expensive and faster but requires more elaborate drive signals.

The basic MOS storage cells are shown in Figure 2.3. A static storage cell using transmission gate selection is shown in Figure 2.3*a*, a dynamic memory cell for use in a random-access memory (RAM) is depicted in Figure 2.3*b*, and Figure 2.3*c* shows a dynamic shift-register stage.

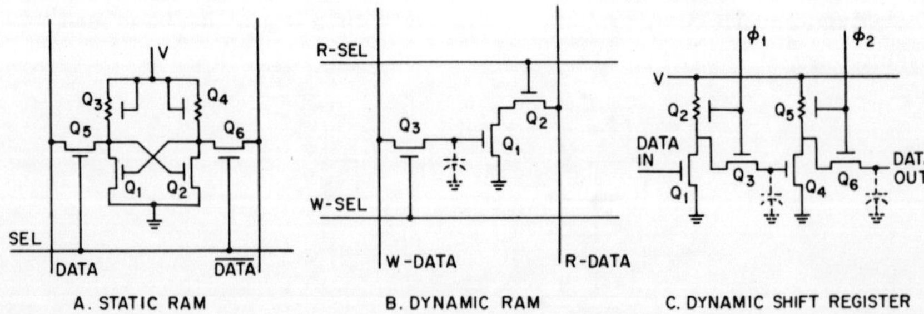

Figure 2.3. MOS storage cells: (*a*) static RAM; (*b*) dynamic RAM; (*c*) dynamic shift register.

In the static RAM stage of Figure 2.3*a*, two static inverters are wired together to make a flip-flop. Devices Q_5 and Q_6 are used as (two-way) transmission gates. When reading, the conducting side of the flip-flop pulls the data line toward ground through these gates. Writing is accomplished by forcing the data lines to the value desired in the cell, thereby overriding the contents of the cell. Because of the small current capability of devices Q_3 and Q_4, it is important that neither data line be near ground when the transmission gates are turned on, or the charge associated with the capacitance of the data line may flip the cell.

A 256-bit static MOS memory chip built using *p*-channel silicon-gate MOS technology is shown in Figure 2.4*a*. The memory is organized as one 256-bit plane, with full address decoding and, except for additional power supplies, it is fully compatible with TTL logic levels. Typical access time is approximately 1 μs. For comparison, a 256-bit bipolar memory is shown in Figure 2.4*b*.

The organization of a 1024-bit dynamic random-access memory chip employing the memory cell of Figure 2.3*b* is shown in Figure 2.5.

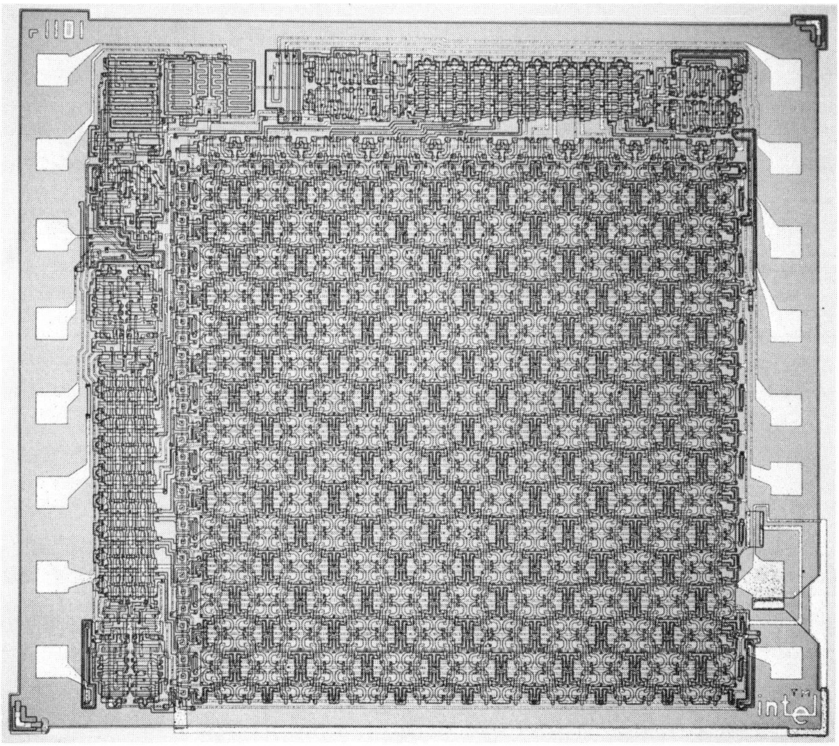

Figure 2.4. Semiconductor memories with 256 bits: (*a*) static MOS, fully decoded, above; (*b*) bipolar, partially decoded, see next page.

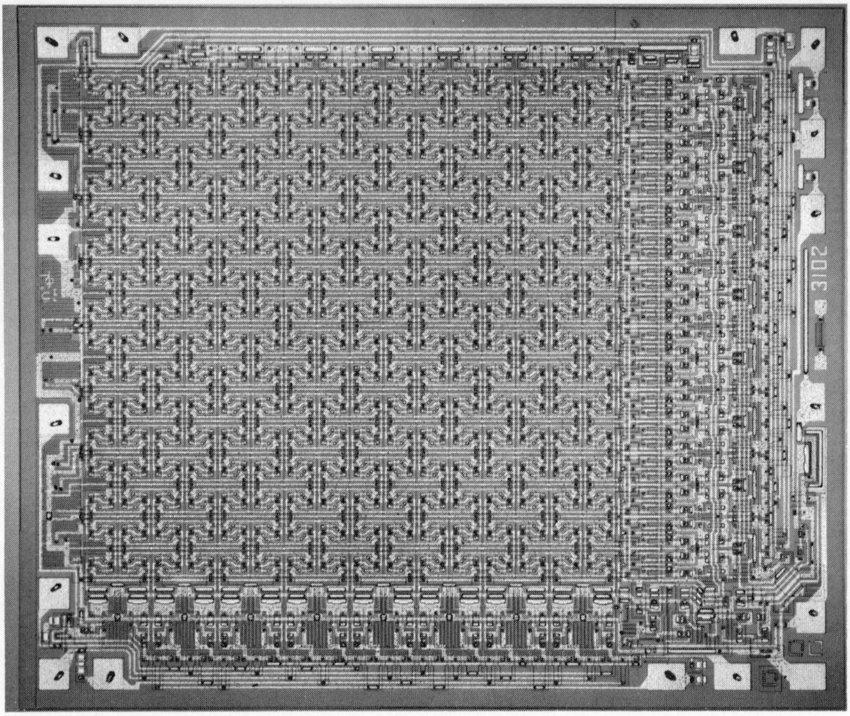

Figure 2.4. (continued)

The input address is fully decoded on the chip. For input–output, a single dual-purpose lead is used. Data from one row of memory cells (as

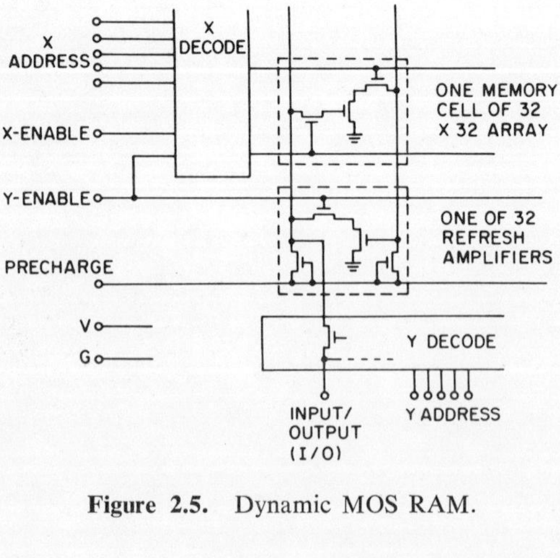

Figure 2.5. Dynamic MOS RAM.

selected by the X address) are transferred to the refresh amplifiers, then the output of one amplifier, selected by the Y address, is read. To write, the contents of the refresh amplifiers are rewritten into the rows of cells, and then the contents of the selected cells are changed by the input data.

It can be seen from Figure 2.3*b* that storage within the cell is achieved by charge storage on the parasitic capacitance of the gate circuit of the storage device (Q_1 in Figure 2.3*b*). The data may be nondestructively read by precharging the R-data lead, then turning on Q_2 with the R-select line. If the gate of Q_1 has been charged, the R-data lead will be discharged. The refresh amplifier inverts the signal on the R-data lead and applies the result to the corresponding W-data lead. This data may then be written into the cell by activating the W-select lead.

In the organization shown in Figure 2.5, reading and writing occur for all cells of one row simultaneously. Because only one bit at a time is available for writing, an (internal) read operation must be performed prior to writing, so that the refresh amplifiers contain data corresponding to the contents of the row into which writing will take place.

There are three clocklike signals associated with the dynamic RAM. These are shown in Figure 2.6. Prior to the start of the cycle, the precharge signal is present (negative) and X and Y enables are absent (at their most positive value).

Upon receipt of the address, the cycle starts by removing the precharge signal, and turning on the X enable. After a specified period of time has elapsed, the Y enable may be turned on. After a specified delay, an output current from the device indicates whether the selected bit contained a "1" or "0." To read this current, the output terminal must be held negative. If writing is to take place, the Y enable is removed while the X enable remains on. The data to be written are then placed on the data input–output lead. To complete the memory cycle, both the X enable and the Y enable must be turned off, and then the precharge signal must be turned on for a minimum period before starting the next cycle.

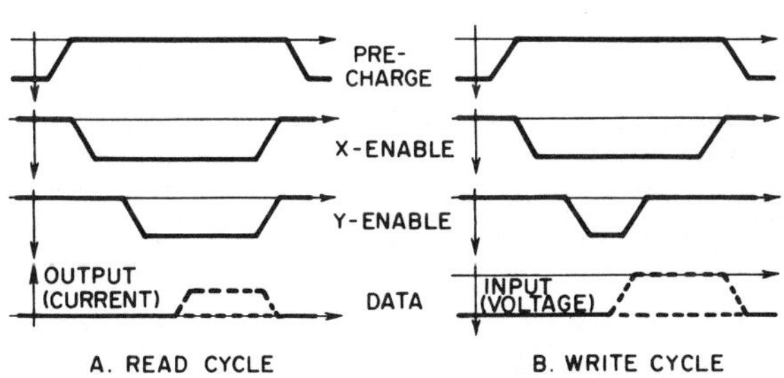

Figure 2.6. Drive signals for dynamic MOS ROM: (*a*) read cycle; (*b*) write cycle.

Both the X and Y enables act as chip selects for both reading and writing. Several chips may have the input–output lead OR-tied to build bit planes of more than 1024 bits.

Using silicon-gate MOS technology, the clock signals and addresses are nominally 20 V peak to peak. Memory cycle times range from 300 to 600 ns, depending upon chip organization and drive signal rise and fall times.

Dynamic MOS shift registers are useful for low-cost serial memories. One type of MOS cell in use is the one shown in Figure 2.3*c*. In this cell, when Φ_1 is on (negative or *p*-channel MOS), Q_2 and Q_3 conduct, charging the gate of Q_4 to the inverse polarity of the data in. When Φ_1 is removed, the charge remains stored at the gate of Q_4. The process is repeated by applying and removing Φ_2. Thus after one clock cycle (one pulse each on Φ_1 and Φ_2), the data appear stored on the parasitic capacitance associated with the output terminal and new data are present at the data input terminal.

A number of variations are now possible for different MOS memory cells. Other MOS phenomena are being studied in an effort to develop new MOS cells.

2.3 READ-ONLY MEMORIES

In addition to the read/write memories described, read-only memories (ROMs) may be manufactured with either MOS or bipolar construction. Most of these memories have read cycle characteristics similar to their read/write counterparts, but the data in the memory are "written" at the time of fabrication. In general, because the memory cells are very small, many more bits of ROM than RAM can be fabricated on a chip of a given size. Bipolar ROMs of 1024 bits (50-ns access) and MOS ROMs of 4096 bits (1 to 2-μs access) are now available. ROMs may be used as substitutes for part of read/write memory and as a substitute for random logic. ROMs are often used to store microprograms for microprogrammed processors.

2.4 SEMICONDUCTOR MEMORY CATEGORIES

The read/write memories to be covered may be lumped into four basic categories: fast RAM, medium-speed RAM, slow RAM, and serial memories. Fast RAMs are used for scratch-pad memories, Caché-type buffer memories, and as arrays of high-speed registers in various parts of a machine. Medium-speed memories may be used for the same applications described for fast RAMs but in lower performance machines. Low-speed RAMs are used in main stores or in peripheral buffers, such as in line printers and card readers. Serial memories are often used for storage in cathode-ray-tube display terminals and they may also be used in other peripheral equipment.

Most of the applications mentioned are relatively conventional, with the memory being organized much like a small ferrite-core memory or, in the case of serial memory, like acoustic delay lines or small drums.

The semiconductor RAMs that have been described may be readily assembled into larger memories, with certain restrictions. These restrictions include a limit on the number of units that may be OR-tied before objectionable loss of speed occurs. Problems associated with the loading of the gates that drive the array may also impose restrictions. Nevertheless, interconnection of the basic RAM packages is quite straightforward. Two problems and their solutions will demonstrate this.

2.5 RAM EXAMPLE ONE

Suppose an application calls for a 256-word, 32-bit/word memory with 16-word by 4-bit/word bipolar memory packages. This requires 128 packages arranged in a 16×8 array. Eight address leads are provided, and each of four of these must drive all 128 packages. The remaining four must be decoded into a 1-out-of-16 code, with each decoded output driving eight packages. Each data input lead must drive 16 package loads and 16 package outputs must be OR-tied for each output. The write gate signal must drive all 128 packages. The lead connections to one memory package are shown in Figure 2.7*a*. A diagram of the full array is given in Figure 2.7*b*.

The full memory system uses, in addition to the 128 memory packages, 16 four-input gates, nine inverters with minimum fanout of 8 or 9, and 40 inverters with a minimum fanout of 16. With typical, off-the-shelf TTL logic, approximately 26 logic packages are required in addition to the memory devices. Typical access times are 70 to 80 ns.

2.6 RAM EXAMPLE TWO

Suppose it is required to construct a 4096-word by 12-bit memory with 1024-by-1 dynamic MOS RAM packages. This calls for 48 packages in a 4×12 array. Twelve address inputs are needed, ten of which may be applied to the packages directly, with the remaining two being decoded and used to gate X enable signals, with each gated X enable signal driving 12 packages. Precharge and Y enable may be delivered to all packages, or may be gated as are the X enables, to reduce power consumption in the memory.

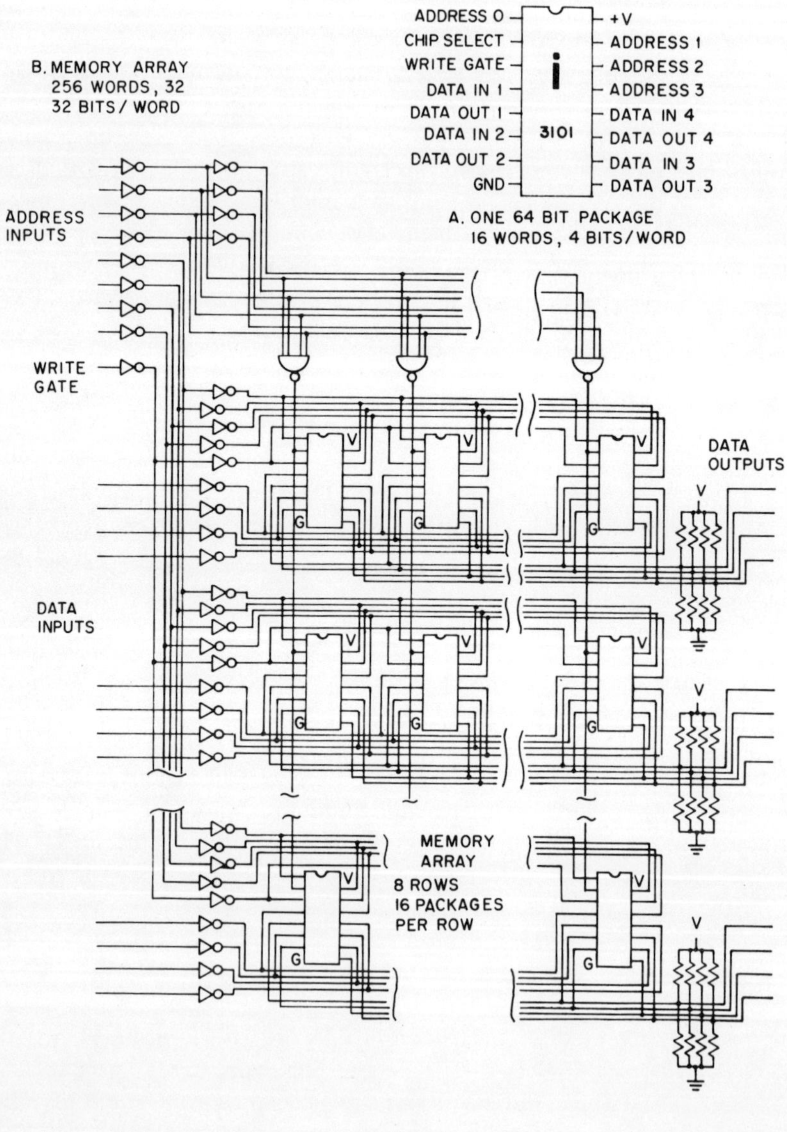

Figure 2.7. Bipolar memory: (*a*) one 64-bit package, 16 words × 4 bits/word; (*b*) memory array, 256 words × 32 bits/word.

For each data input–output signal, the I/O leads of four packages must be OR-tied. A special word driver–sense amplifier must be provided for each I/O lead. A suitable circuit is shown in Figure 2.8*b*. One package of memory is shown in Figure 2.8*a*. The memory array is depicted in Figure 2.9. The level changers (blocks labeled *L*) are circuits that convert from logic levels to 20-V peak-to-peak signals. Each TTL circuit must drive approximately 500 pF. The blocks labeled *W/S* are the word driver–sense amplifiers. In this array, the terminal labeled *G* in Figure 2.5 is connected to +20 V, and the terminal labeled *V* is connected to ground.

2.7 REFRESH ACTION

The memory array shown in Figure 2.9 must be driven by a control circuit that includes provision for refresh action. Because the storage in the memory cells is dynamic, the data must be periodically read from the cells and rewritten. With the chip organization shown in Figure 2.5, executing a read/refresh cycle restores the data in one row of cells. To refresh the entire chip, 32 refresh cycles must be executed; i.e., one for each of the 32 possible *X*-address states.

To accomplish refresh action, a refresh controller may be constructed and attached to the memory as a high-priority cycle-stealing device. Since the longest time a cell may retain data without refreshing is approximately 2 ms, the cycle-stealing refresh control should steal one cycle every 3/32 ms, or approximately one cycle every 60 μs. The stolen refresh cycles operate the *X*-address states in sequence, so that in any 2-ms period, all cells are refreshed. In order for the entire memory to be refreshed, the decoding associated with the *X* enables (and precharge, etc.) is overridden. Data on the data lines are ignored during these refresh cycles.

Refresh cycle control need not be provided if the memory is operated sequentially with an adequately high cycle rate. Other refresh control techniques may also be used. For example, separate circuits, either analog or digital, may be provided to determine which *X*-address rows have not been refreshed by normal operation. In general, this technique is more

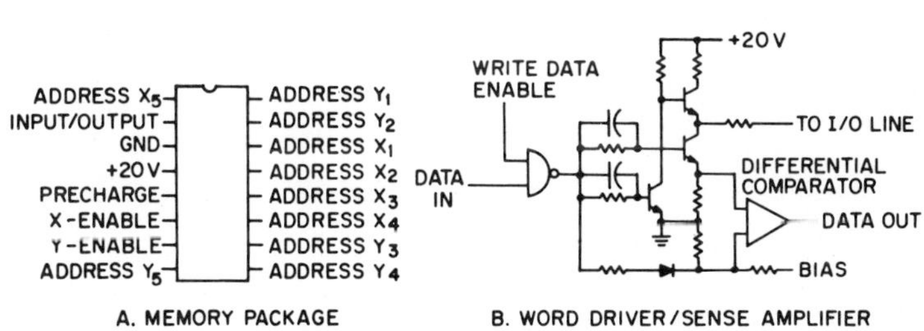

Figure 2.8. Components for a dynamic MOS RAM: (*a*) memory package; (*b*) word driver–sense amplifier.

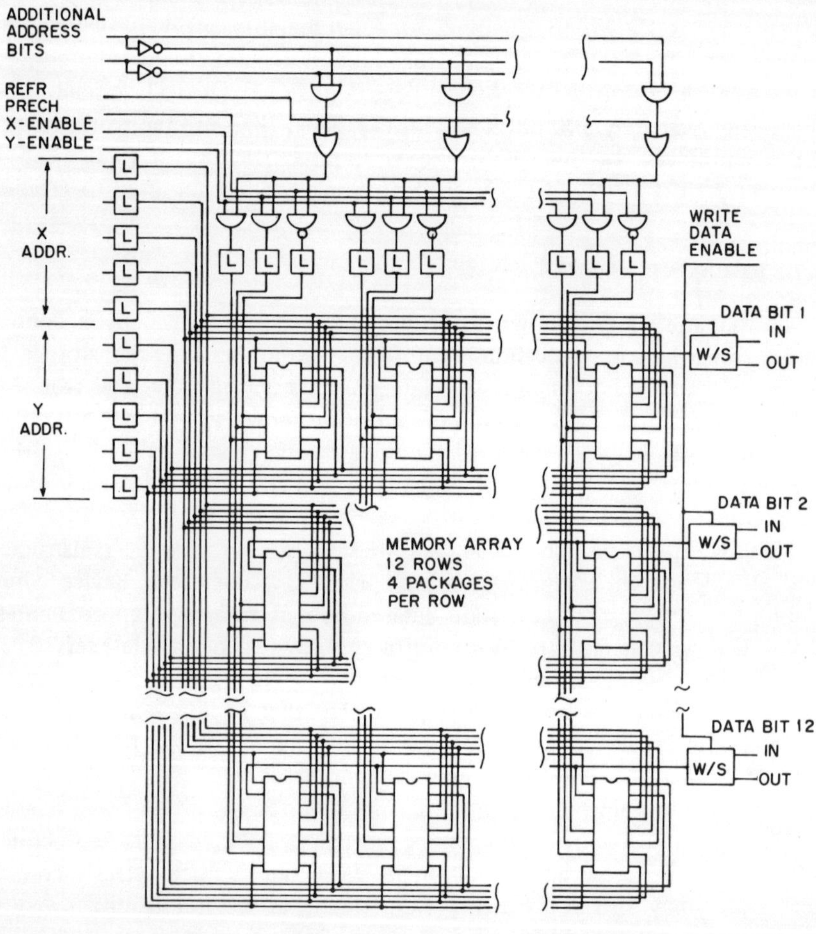

Figure 2.9. Memory array with 4096 words, 12 bits/word; dynamic MOS RAMs with 1024 bits used.

expensive. Even if all refresh cycles can be saved, the total saved represents only 1% of the available memory time.

Shift-register memories are relatively simple to drive. The primary consideration is clock-driver current capability. For silicon-gate technology, the loading of each clock driver is approximately 0.2 pF per bit. Thus a 100,000-bit memory would load each clock driver with 20,000 pF. If 15-V clock pulses with 50-ns transitions are required, each clock driver must supply peak currents of approximately 6 A. In general, multiple clock-driver circuits must be used to provide this total capability. The memory may be divided into sections with little or no change in cost. At the expense of a small amount of gating, connection of the shift registers in series may be replaced by a parallel structure with random selection of sections but serial access to data with a section.

The shift register performs much like a very fast drum memory. Unlike the drum memory, however, the shift register may be instantaneously "stopped" or started, subject to the restriction that the memory must not remain stopped for more than a given time, usually 1 ms.

2.8 VOLATILITY CONSIDERATIONS

All the semiconductor memories now available, with the exception of ROMs, are volatile storage devices. For these devices to retain stored data, some power must be applied. Static MOS and certain bipolar memories may have provision for reduced power operation in which no read or write operations may take place, but data are not lost. In general, to retain data the power levels required are in the order of several hundred microwatts per bit. However, dynamic MOS memories require a considerably smaller amount of lower data-retention power.

Each 1024-bit package of a dynamic MOS RAM has a total capacitance (sum of all input capacitances) of approximately 120 pF. At most, one 20-V transition in each direction must be made per input per memory cycle. To retain the data, 32 memory cycles must be executed in each 2-ms period. To charge the 120-pF capacitance to 20 V, 2400 pC must be delivered by the power supply. Since only one transition draws power from the supply, the approximately 16,000 memory cycles per second draw a total charge of about 38.4 μC from the supply per second. The total power to execute these drive signals averages under 800 μW, or 0.8 μW per bit. The dc power used by the cells is a fraction of the clock power, and an allowance must be made for the clock-driver bias currents. Nevertheless, using efficient clock drivers, the total power requirement to retain data should be in the range of 1 to 2 W per million bits. At these power levels, even a small battery pack may be sufficient for retention of data for several hours or more.

Similar calculations may be made for shift-register memories. Assuming 15-V clock pulses, 0.2 pF per bit, and 1-ms maximum retention time, the power drawn from the power supply to drive the two clocks amounts to approximately 0.1 μW per bit. The dc power consumption in the clock drivers and the shift registers themselves may bring the total power to about 0.25 W per million bits, or approximately an order of magnitude better than the dynamic MOS RAM. For either type of memory, a modest battery pack and a suitable controller may render the memory effectively nonvolatile.

2.9 SYSTEMS CONSIDERATION

Most of the solid-state memory products that have been described can be used in a wide variety of configurations. In most cases, the cost of the memory system is nearly independent of the way the memory is organized, depending instead on the type of solid-state memory product used and the total number of bits used. In general, this relationship permits greater parallelism of data access than is available in magnetic memories. For example, a 256-word, 32-bit memory will require approximately the same numbers of peripheral gates if organized as a 16-word by 512-bit/word memory. In the case of a dynamic dual 100-bit MOS RAM, a reorganization of the packages to give wider words would require additional word-line–sense driver circuits. Even at maximum parallelism (1*k* words of 48 bits each), the cost of the additional word drivers adds less than 30% to the cost of the memory.

The high degree of parallelism possible with large semiconductor memories presents a challenge to the systems designer. Many instructions or large blocks of data may be simultaneously accessed. To use the memory most efficiently, means must be developed to cope with these parallel data.

Not only may a high degree of parallelism be obtained with semiconductor memories, but several types of memory components may be mixed on a single printed-circuit card. For example, MOS shift registers may be combined with bipolar RAMs. Suppose 1024 packages, each containing a 1024-bit-long shift register, are combined with 256 packages of bipolar RAMs, each organized as 16 four-bit words; transfers may take place 1024 bits at a time. If a 32-bit machine is used, and the shift registers are capable of operating at 1 MHz, a transfer rate of 32 words per microsecond is possible. The 512-word contents of the bipolar memory can be loaded or unloaded in 16 μs. Such a bipolar buffer might be used as a Caché-type buffer for the serial memory, or as a multiplexed interface buffer for a medium-size dynamic MOS RAM, the serial memory, and a central processor.

2.10 PROCESSOR LIMITS

A primary condition for eliminating the need for the backup (dynamic MOS) RAM, is the ability to keep the central processor sufficiently busy during the access times to the serial memory. Several characteristics of existing program encoding techniques reduce this possibility, however. These are:

1. Data and program codes are often widely separated. In particular, if many data values are used, these may be scattered over a wide range of memory.
2. A program may have to wait for data that are not in either memory. Interactive programs may require human intervention, and some programs may use too large a data base to fit within the active memory.
3. No estimate of activity of a given amount of program code is made available to the machine or compiler. Thus, even in a multiprogram execution environment, the central processor is offered no information on when a program will require access to presently unavailable data or program.
4. The existence of conditional jumps within a program reduces the ability to predict when additional program and data must be made available.

These factors may be even more evident in a processor using a wide word because of the potentially higher processing rates. By providing access to the full 1024 bits available from the memory described, some 32 instructions are available at intervals of less than 100 ns. Most processors execute instructions in sequence, and many do not have the capability for executing even a single instruction in one cycle of a memory of the speed described. With parallel processors, much higher rates are potentially available, but most compilers do not distinguish between parts of a program that may be simultaneously executed and parts that must be sequentially executed. Meanwhile, if higher rates are achieved, the problem of keeping the central processor and its associated fast memory busy will become even more difficult.

Several of these problems have become apparent in high-performance processors. Large amounts of time may be wasted waiting for data to be made available to the processor. Some of the means to reduce this waste include the use of multiprogramming or time sharing. To permit active residency of sufficient program material, very large random access ferrite-core storage units (in the order of 10^7 bits) have been provided. These large storage units have usually had significantly lower speed capability than the main computer memory.

2.11 IMPROVING PROCESSOR PERFORMANCE

In a ferrite-core memory system, the memory has tended to place limits on the system, as well as the peripherals. The use of Caché-type buffers has helped to remove some memory-introduced limitations. However, with more extensive use of semiconductor-memory devices, the possibility exists that the central processor and the peripheral devices will constitute the major source of system performance limitations. Some of these limitations might be reduced in several ways. Techniques to improve processor performance include parallel processing and pipeline organization. To reduce limitations associated with peripheral limitations, larger systems may be used to permit a wider variety of jobs to be simultaneously active, or look-ahead techniques may be used. The use of a hierarchy of memories with separate memory management processors all performing look-ahead and data reorganization is another possible approach to this problem.

Although pipeline techniques have already been used in high-performance general-purpose computers, parallel computation has tended to be limited to more specialized machines or has been very limited in scope. For example, associative processors and machines such as Illiac IV have required parallel data (data arrays) to show a speed advantage over a conventionally organized computer. Nevertheless, the serial nature of most program codes currently written and executed is seldom fundamental to the problem. However, the form parallelism may take is a highly variable function of the nature of the part of a program being considered. A great percentage of this parallelism may be determined at the time of compilation, and if a computer is suitably organized, several processing elements can be kept busy, using parallelism information provided by the compiler and included with the program code.

To reduce the wide range of access needed for data, separate data and program memories might be used. (Many program systems do not make extensive use of processor-executed program modification.) By including additional data transfer instructions in a program at compile time, or with the use of separate memory processors, needed data can be made ready in a small fast memory in time for the main program to operate with this data. This pseudo-look-ahead technique might be accomplished with the memory processor acting in much the way a conventional I/O channel does today. Status bits associated with the data words may be used to prevent erroneous data manipulation, such as attempting to use data from a storage location before the data have been loaded from the larger, but slower, main memory or bulk store.

Look-ahead techniques have been used in some of the highest performance computers. Even more effective use may be made in a memory hierarchy if each memory includes either automatic look-ahead, or compiler-generated look-ahead information.

These techniques are a few that may be developed in efforts to increase the speed and processing capability of general-purpose computers and to take advantage of some of the potential of semiconductor memory devices.

Many of the changes proposed may require greater knowledge of compilation and execution of programs. Although many programs do not readily lend themselves to look-ahead or parallel processing, a mix of programs may provide a reasonable base for this type of operation. It is likely that some analysis of existing programs will be necessary to determine where the greatest promise lies. It may also be necessary to develop new languages to make best use of new processor organizations.

The use of multiple processors and separate memory management processors may be expected to increase the cost of the central processor portion of a computer system. These cost increases should be offset by the trend toward reduced central processor cost produced by low-cost semiconductor devices and integrated circuits. To the extent that elements of these processors can be standardized, cost savings may be obtained by the use of large volumes of standard large-scale integration (LSI) components.

2.12 THE FUTURE

Questions have been raised within the semiconductor industry about the need for hierarchies of memory. For the foreseeable future, significant cost differences will exist between different types of memory, with serial MOS memories being the least expensive and fast, bipolar RAMs being the most expensive. Larger memories tend to be slower owing to the relation between fanout and delay, which in turn is related to the gain–bandwidth product of available semiconductor devices. Therefore, even when only one type of memory is used, distributing the memory in smaller parcels or organizing the memory as two parts of different size may offer performance advantages.

A wide variety of semiconductor-memory devices will soon be available to the computer manufacturer. Although these devices will most likely be used for direct replacement of core memories, the semiconductor memory promises eventually to offer much higher performance at lower cost. To take full advantage of semiconductor-memory potential, the system designer must seriously examine existing techniques to see how the unique characteristics of new devices may be put to best use. The system organizations that have developed around the ferrite-core memory may not necessarily be the best for use with semiconductor memory.

3

Semiconductor Memories: Make or Buy?

by Robert Norman

Nortec Corporation

The decision to make or buy is by definition an economic one. The make-or-buy choices concerning memories that are made today will affect the expenditure of large amounts of money over the coming years.

Five years ago many systems architects were pleading for a breather in technological change so that decisions being made then would not be obsolete by the time they were implemented. Instead, large-scale integration (LSI) was invented.

Semiconductor memory is now receiving the concerted attention of highly skilled technologists steeped in IC know-how. Considering that a memory is a well-defined, closed system relative to the other problems that the same technologists have attacked in the past, a significant rate of growth in semiconductor-memory technology can be expected. It is conceivable that all current concepts of the architecture of data-handling systems will be obsolete in five years.

The systems company decision maker finds himself making choices affecting his position in a market in motion and dependent on source technology, which is also in motion. Decisions made within these fluid constraints must be consistent with each company's resources, goals, and aspirations. The engineer today must take this fluid technology and put it to work to enhance his company's position in the exploding marketplace in a manner commensurate with the resources and the profitability and return-on-investment aspirations of his company.

3.1 WHO BUYS MEMORY?

In the final analysis it is the end user of a data-processing service who buys memory. He is not concerned with whether the memory is core or semiconductor or ants. He is not concerned with the question of MOS versus bipolar. His concern can be simply stated: He requires that the value of his data-processing system be greater than its cost. The data-handling market is exploding because system costs have been reduced below the value of the systems to the end user.

3.2 WHO SELLS MEMORY?

Memory is sold ultimately by the system supplied who installs data-processing systems for the end user. He may purchase and lease out data-processing systems, assemble them for end users, or manufacture them; or he may manufacture some portion of the end-item hardware and purchase the rest. As he uses the hierarchy of input–output devices and a hierarchy of logic and memory distribution among them, he also uses a hierarchy of memory—all aimed at providing maximum values to the end user at minimum cost.

It is inconceivable that modern cost-effective systems would make no use of magnetic memory. It is just as inconceivable that modern cost-effective systems would make no use of semiconductor memory. Today's system supplier must provide a hardward-and-software configuration that is adequate for the needs of the end user at minimum cost. This will involve a mix of memory types.

3.3 WHO BUYS SEMICONDUCTOR MEMORY?

The company that buys semiconductor memory does so to fill some part of its total memory requirement based on the characteristics of performance, flexibility, and cost of semiconductor memory relative to other techniques, and not because it is popular today or because of its romantic appeal. The buyers of semiconductor memory are: (1) systems manufacturers, suppliers, or firms that do both, (2) computer manufacturers, (3) peripheral-equipment manufacturers, and (4) memory manufacturers.

The nonmanufacturing systems supplier will most likely buy semiconductor memory as a complete memory unit that can be simply attached to his other purchased systems components. The memory manufacturer, on the other hand, will most likely buy only wafers, chips, or packaged dice. The remaining companies may buy semiconductor memory in any form or any combination of forms.

3.4 HOW CAN SEMICONDUCTOR MEMORY BE BOUGHT?

Semiconductor memory can be bought in almost any form. For example the memory user can purchase: (1) unprocessed silicon; (2) processed wafers, dice, packaged devices at vendor risk; (3) processed wafers, dice, packaged devices at customer risk; (4) assembled memory modules/ subsystems; or (5) packaged memory systems.

Each of these alternatives can be evaluated in terms of several considerations of interest to the buyer. As part of the assessment, time must be taken into account. Solutions that seem best for the short term may not be best for the long haul, and vice versa. The important business considerations are the effects of the decision on the profitability and return-on-investment characteristics of the company and its available resources. The technological considerations are flexibility in the application of the memory and insensitivity to technological obsolescence.

Two considerations of the future that affect decisions made about memory today are: (1) semiconductor memory will be manufactured, sold, and used in very large quantities and (2) only the LSI configuration can easily be a "standard" product.

As a consequence, the semiconductor memory will represent a major source of revenue to major semiconductor manufacturers and to that newly emerging segment of the semiconductor industry specializing in memory alone. The effect of these considerations is shown in the price projections of Figure 3.1. These projections indicate a continuous decrease in the cost of fabricating semiconductor memory throughout this decade. It is expected that the cost will be less than the price of medium-size core memories at some time before the end of 1971.

The price of semiconductor memory should decrease until it becomes comparable to the price of cores; then there will be no further reason for reduced prices until about the middle of the decade. Then, even though memory use will be very large, it will be matched by memory fabrication capacity. At that time competition among the fabricators will force the price down toward the semiconductor-memory cost level. The price could drop earlier because of well-known pricing practices of the semiconductor industry. There reductions will be achieved through continuous technology and technique improvements aimed at cost reduction.

It is now appropriate to examine the alternative ways of purchasing semiconductor memory.

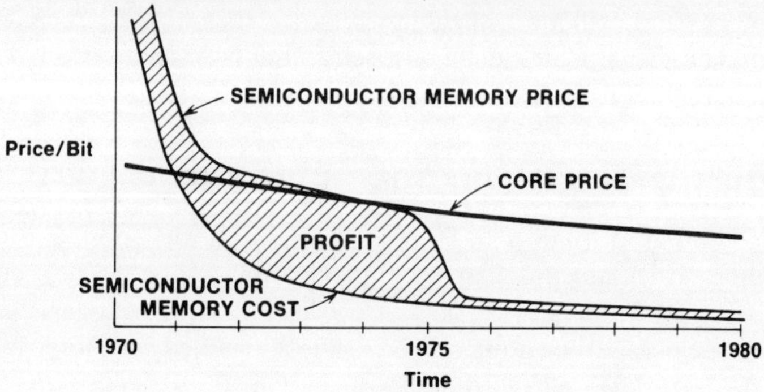

(a)

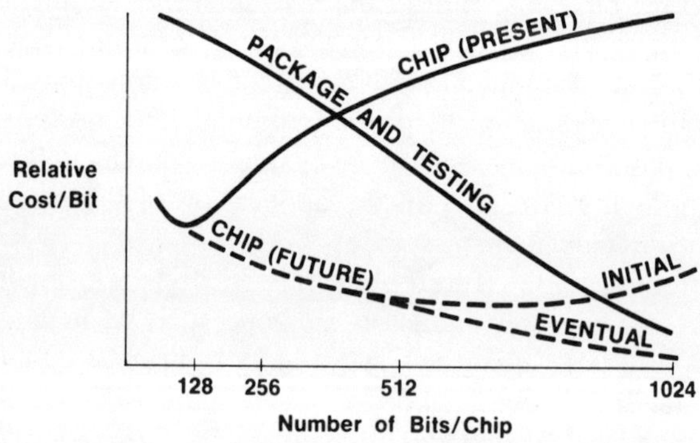

(b)

Figure 3.1. Price-per-bit curve for semiconductor memories will eventually approach cost curve. Present relation between chip cost and packaging–testing costs will change.

3.5 UNPROCESSED SILICON

Companies that "purchase" memory as silicon already have an operating semiconductor fabrication plant as part of their capability. On the surface, by operating in this fashion, they have absolute control over the design of the memory and can therefore optimize it for their own use. They are able to incorporate new technology at will for any necessary improvements of memory performance or for reducing cost. They also have the highest value-added approach, since the cost of materials for memory with good yield is about two mils per bit. As shown in Figure 3.1, during the first half of this decade, for this type of memory-using company, the investment can be modest and the return on investment good.

Why doesn't everybody take this approach? One reason is that the supply of talent capable of operating a semiconductor plant effectively is critically short. The data-processing-systems manufacturer may well be able to come up with the several million dollars required to get into the semiconductor-memory business, but he may not be able to discriminate against his existing staff sufficiently to recruit and retain the key personnel needed to man a semiconductor facility.

Risks that must also be considered when contemplating installation of a semiconductor operation are: (1) enormous operating running losses, (2) inability to implement any technology beyond that with which the facility starts, and (3) frozen memory designs resorted to in order to achieve "manufacturability" of the memory. If problems occur, it is also possible that inside customers will shift to outside sources.

3.6 WAFERS, CHIPS, OR PACKAGED DEVICES (VENDOR RISK)

Semiconductor memory may be purchased as finished devices on wafers or separated into chips or in packages. With standard products, the vendor assumes the risk for the parts he sells; that is, the customer pays for good devices. Over the long haul it can be expected that memory purchased this way will be highly competitive in price and performance. Usually more than one source for parts will be available. Proper exercise of the purchasing function by the customer will permit the highest degree of assurance of having an adequate supply of parts for his data-processing systems, at a level still permitting room for value added in assembling the parts in memory subsystems.

The penalties attached to buying memory this way are: (1) performance tends to lag one to two years behind the state of the art, (2) prices could stay at core levels for a few years as mentioned earlier (this may not occur, since the semiconductor industry sometimes lowers prices even when it is not supplying the available market), and (3) memory subsystem packages will be subject to continuous redesign to accommodate a continuously increasing number of bits per device (through which lower costs will be achieved by the semiconductor manufacturer).

Companies that buy wafers or chips from semiconductor suppliers and do their own packaging may not achieve significant value added because memory dice will always be relatively expensive. Reduced bit costs will be achieved principally in higher bit densities, thus maintaining the price level of the finished device while reducing the price per bit.

3.7 WAFERS, CHIPS, OR PACKAGED DEVICES (CUSTOMER RISK)

The time required for a product incorporating new memory technology to appear on the market can be shortened by purchasing memory as wafers, chips, or packaged devices but fabricated to custom design by a semiconductor processing facility. It is possible to anticipate competitive commercially available memory one to two years downstream and to incorporate such memories at a "competitive" price in current equipment designs. This permits significant value added during this interval, and memory cost that tracks the memory cost curve shown in Figure 3.1 rather than the price curve. The risks, of course, are well spelled out—the design is the customer's. The customer by definition will tend to freeze the technology he is purchasing; but once again, he may be one to two years ahead of commercially available technology.

3.8 ASSEMBLED MEMORY MODULES/SUBSYSTEMS

Several semiconductor companies have been established which supply memories in assembled memory modules/subsystems form. When memory is purchased this way it is not necessary to redesign the system memory every time a new device appears on the market. The vendors in this case normally sell memory in larger blocks and make their decisions based on optimum number of bits per chip and on-chip or off-chip locations for decode, drive, and sense circuitry. Such companies make available standard blocks of memory with relatively stable functional characteristics, although performance and price improve over the years. The price a customer pays for this, of course, is low value added and minimum flexibility in fitting the memory both functionally and physically into the system.

3.9 MEMORY SYSTEMS

Packaged memory devices are being bought in large quantities by customers who would not have considered buying core planes simply because memory devices are compatible with their circuitry and the development and manufacture of memory subsystems is far simpler than the comparable task of using cores. At the other end of the scale, some major equipment manufacturers who commonly have their own core plane and memory subsystem fabrication plants also buy semiconductor memory as processed devices, wisely avoiding the pitfalls of entering the semiconductor business.

4

Semiconductor Memories: Current-Mode Logic versus Other Approaches

by Andrew R. Berding and Robert C. Lutz

Advanced Memory Systems, Inc.

The designer of a high-performance semiconductor-memory system has many techniques from which to choose. Circuit approaches available to him include transistor–transistor logic (TTL), Schottky-diode-clamped transistor–transistor logic, hybrid MOS–bipolar logic, and current-mode logic (CML). He can choose functional organization, where each chip acts as a self-contained memory, or nonfunctional organization, where decode and sense functions are performed off the chip. He can choose circuits that are saturating, nonsaturating, or clamped. And he can choose bit organization or word organization. The designer must decide which of these approaches provides the best performance, cost–power dissipation, and reliability for his application.

4.1 PERFORMANCE

From the standpoint of system speed, the choice is likely to be CML. Current-mode logic provides the fastest speeds, followed by clamped TTL, saturating TTL, and MOS–bipolar hybrid. These speeds reflect the best commercially available examples of each of these approaches.

It may be clear that current–mode logic is faster than the other approaches because it is nonsaturating, but why it is so many times faster is less obvious. Its faster speed results in part from its greater logic flexibility—with CML, both the normal and inverted outputs are available at no extra delay, wired emitter ORs are free, wired collector ANDs are possible, and multiple outputs can be used. All these logic functions can be obtained in one logic stage delay. This means that the same functions can be accomplished with fewer stages of logic than are required with TTL. And since each of these stages also happens to be much faster than a logic stage in either saturated or clamped TTL, the chip speeds will be far greater.

The decoder used on a commercially available clamped TTL memory chip is shown in Figure 4.1. Notice that to generate both phases of the address line requires another stage of delay. In addition, each line has a heavy nine-way fanout. In contrast, Figure 4.2 shows how the same decoder function can be implemented on a CML memory chip. Not only does the CML input circuit buffer the input signal, but it produces both phases and

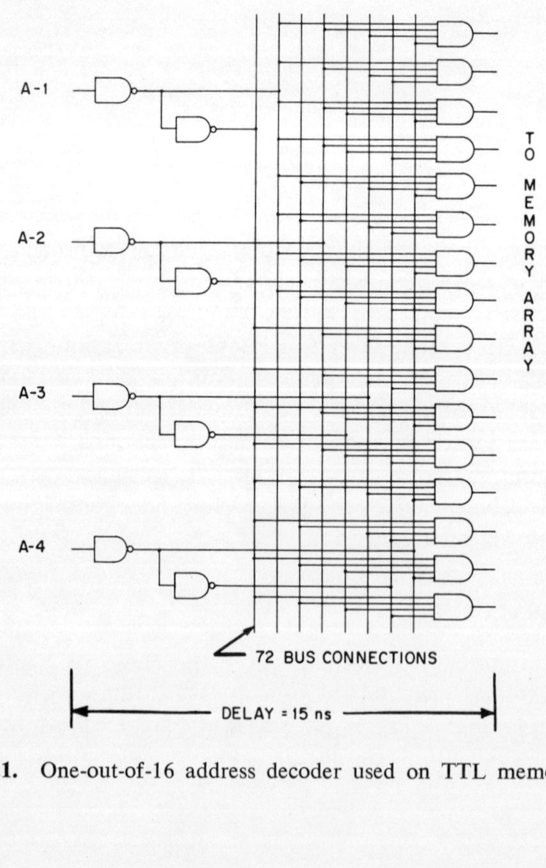

Figure 4.1. One-out-of-16 address decoder used on TTL memory chip.

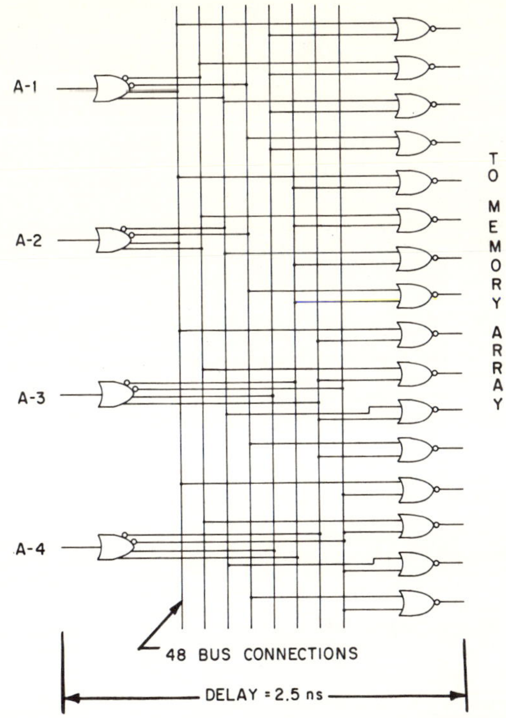

Figure 4.2. One-out-of-16 address decoder used on CML memory chip.

performs a 1-out-of-4 partial decode. This total function is accomplished in only one logic delay—approximately 1 ns. In addition, the circuit has light four-way fanout. The result is a large difference in decode time—2.5 versus 15 ns. Also, although it may not be self-evident, the CML circuit is much easier to lay out and is more compact than the TTL version, since it uses only two input ANDs instead of four input ANDs and its wiring topology is simpler.

Because of its logic flexibility, the CML approach allows the important functions of column selection, sense amplification, and data ORing to be performed all in one stage delay. Since CML allows the memory support functions to be achieved efficiently, the access through the fully functional CML memory chip is achieved in only four logic delays (approximately 5 ns). This delay time includes input buffering, generation of both phases, full decoding of cell information, sensing, ORing, and output buffering.

Another advantage of CML memory is that sensing and writing occur at a low-impedance mode. This virtually eliminates write-recovery time, which greatly enhances cycle time. Write times of less than 2.4 ns are typical. This is how 10-ns cycle times are achieved.

4.2 COST

The high layout efficiency of CML memory with on-chip support circuits results in more dice per wafer. This more than offsets the slightly higher wafer processing cost and the differences in yield. CML memory chips are built with normal manufacturing tolerances such as 0.2 mil (5 microns).

Cost and performance are two factors encouraging the organization of a chip on a functional basis rather than on a nonfunctional basis—that is, letting each chip act like a self-contained memory with input buffering, decoding, cells, sense amplifiers, output powering, and write-control circuits on the chip. Cost is reduced because there is much less silicon area allocated to pads, there are fewer wire bonds, and low-cost industry standard packages can be used. Performance is enhanced because the drivers, cells, and sense amplifiers are all on the same chip and their transistor parameters all track. Furthermore, the I/O interconnection capacitance, which severely degrades performance, has been eliminated and the noise susceptibility of the memory improved. Reliability, which is very dependent on both the number of interconnections and the noise susceptibility, should also be improved.

4.3 RELIABILITY

A better understanding of system reliability can be obtained by comparing bit organization with word organization. With bit organization, only one bit per word can occur on a chip or card. With word organization, more than one bit of the word and perhaps the whole word can be handled on a chip or card.

A word-organized chip requires more I/O interconnections. Figure 4.3 shows how the number of interconnections grows as the organization is changed to decrease the number of words while increasing the number of bits per word. Since the number of storage chips required to make the complete system is constant, the number of interconnections in the system will grow proportionately, and thus system reliability will suffer. Even worse, the failure of any chip will cause more than one bit per word to fail, so that failures can go undetected even with parity error detection. However, with bit organization, although a chip or card failure can cause the loss of a bit in many different words, only one bit per word will be lost, and this can be detected with a simple parity check or corrected with a single error-correcting Hamming redundancy code. Thus bit organization will not only reduce the number of failures, it will also reduce the consequences of a failure.

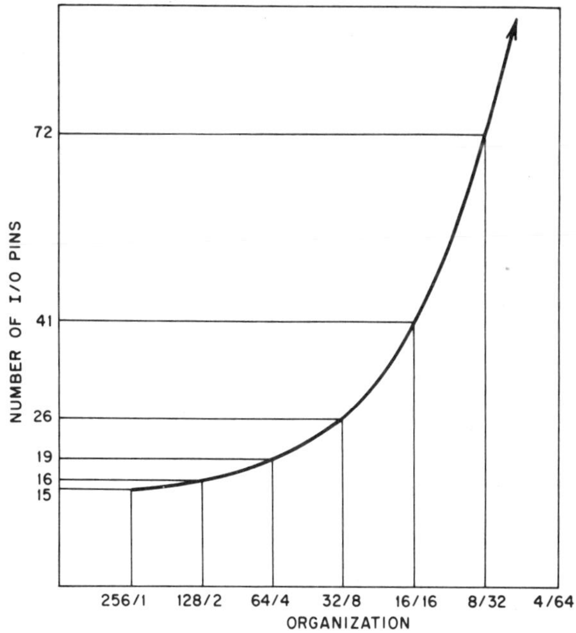

Figure 4.3. Increase in interconnections as organization is changed to decrease the number of words while increasing the number of bits per word.

4.4 POWER

Although it is sometimes four to five times faster than clamped transistor–transistor logic, the CML memory dissipates less power. Again, the logic efficiency of CML allows many fewer circuits to accomplish the same function; therefore less power is required. The transistors are operating in a much faster mode (emitter followers and grounded base rather than grounded emitter as with TTL), so less power is needed to get the required speed. Perhaps the example chosen is somewhat unfair because the clamped-TTL memory chip used is word organized, which means that it is necessary to repeat the very high power output buffer and write circuitry four times. If instead that chip were bit organized, the power dissipation would be comparable, but not the performance. MOS–bipolar memory, with its inherently lower standby power, dissipates considerably less power than do other approaches. Thus for power-critical applications, MOS–bipolar hybrid memory may be the best choice in spite of its other limitations. It is interesting, though, that CML memory has a better speed–power product than even the MOS–bipolar approach.

5

Ion-Implanted Read-Only Memories

by Carroll R. Perkins, H. G. Dill, and T. N. Toombs

Hughes Aircraft Company

The advent of ion implantation techniques [1–4] has allowed the development of large-scale integrated read-only memories with cycle times of less than 300 ns and power dissipation of less than 400 mW for fully decoded arrays. Furthermore, ion implantation offers another method for permanently storing binary word information into the MOS array.

5.1 ION IMPLANTATION

Ion implantation processing results in gate-to-source and gate-to-drain overlap capacitances that are typically one-fortieth of the capacitances associated with standard *p*-channel MOS processing techniques and approximately one-tenth of the overlap capacitance associated with other self-registering gate technologies, such as silicon-gate technology. Ion implantation appears to be the ultimate technology for gate self-registration. A better understanding of how its low capacitances occur can be obtained with the aid of Figure 5.1, which shows cross-sectional views of standard *p*-channel, self-registering gate, and ion-implantation processes.

A cross-sectional view of the standard *p*-channel MOS device appears in Figure 5.1*a*. The typical dimensions given in this cross-sectional view depend on two process factors. The first is a function of the lateral or

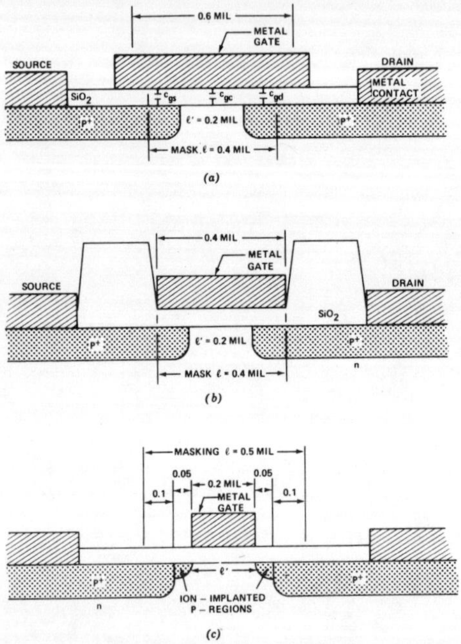

Figure 5.1. Cross-sectional views of various *p*-channel MOS devices: (*a*) standard; (*b*) self-registered gate structure; (*c*) ion-implanted structure.

sideways diffusion of the *p*-type diffusion into the *n*-type substrate, and the second depends on the location of the gate metalization. The diffusion depth is controlled to obtain appropriate breakdown voltages for specific applications. As diffusion depth is increased to improve the breakdown voltage characteristic, common diffusion technology causes the lateral diffusion to be equal to the diffusion depth; thus, a 3-micron diffusion depth results in a 3-micron sideways diffusion that will vary from manufacturing lot to manufacturing lot. With the variation of mask tolerances and alignment tolerances associated with standard semiconductor photolithographic processes, the gate mask must incorporate enough tolerance to assure that the gate protrudes over both the drain and source diffusions in order to produce MOS devices with high yields. The dimensions specified in Figure 5.1*a* are typical for the standard *p*-channel process and represent a means of comparing other MOS processes for the all-important capacitance parameters.

By way of comparison, Figure 5.1*b* shows a cross-section view of the typical self-registered gate structure. The self-registered gate process eliminates the need for aligning the gate mask after the *p*-type diffusion, since the drain-and-source diffusion is accomplished in such a manner as to accurately define the gate area. Although self-registration removes the variable associated with gate alignment, it does not eliminate the overlap capacitance associated with the sideways diffusion. Thus, the typical dimensions shown in Figure 5.1*b* reveal that the overlap capacitance has been reduced by a factor of 4 from that of the standard *p*-channel process.

5.2 ELIMINATION OF PROCESS VARIABLES

To eliminate both the sideways diffusion and gate location process variables, ion implantation is used on wafers that have been completely processed through the metalization step. Ion implantation occurs in areas under thin oxide only. Metalization, thick oxide, or photoresist inhibits implantation. Thus, the gate metal defines very accurately the channel length, and the possibility of sideways diffusion capacitance is eliminated as a variable. Typical fabrication dimensions are shown in Figure 5.1*c*. The overlap capacitance for ion-implanted MOS arrays is virtually 0 and has been empirically measured as approximately one-fortieth of the overlap capacitance of the standard *p*-channel processes.

This huge reduction in overlap capacitance has the net effect of reducing the Miller capacitance in an integrated MOS array by a factor of about 5. The increased gain–bandwidth characteristics allow operating speeds of MOS circuitry to increase typically by a factor of 5. By way of an example, Figure 5.2 illustrates an ion-implanted dual 64-bit dynamic shift register, which is capable of handling data rates in excess of 10 MHz. Another unique aspect of ion implantation is its ability to produce high-value resistors in a very small area. Note that the S-shaped area in each shift-register cell of Figure 5.2 represents a resistor of approximately 140 kΩ This resistor is contained in an area approximately 2 by 3 mils. The presence of this linear resistor allows internal nodes to be pulled up to the supply voltage (V_{DD}) without the need for a second voltage supply (V_{GG}). Furthermore, for the same speed, the power dissipation of the implanted resistor is about one-half that of a conventional MOS device.

The sheet resistance of the ion-implanted regions can be controlled to have values in the range of 1 to 10 kΩ per square. Furthermore, ion implantation requires an extremely shallow diffusion, with resulting low capacitance per unit area. The ability to implant precision resistors has already allowed the manufacture of monolithic, precision 9-bit digital-to-analog converters where the precision resistor ladder network is ion implanted in the basic MOS array.

Ion implantation is not restricted to boron diffusion into silicon [5], nor to *p*-channel technology. Work has been conducted on developing reproducible systems for implanting both phosphorus and boron into comple-

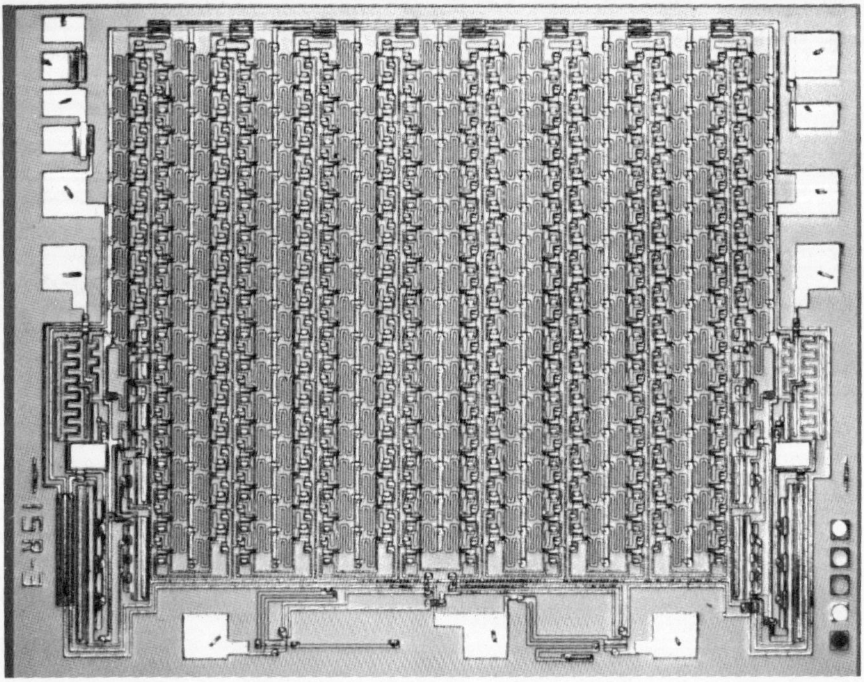

Figure 5.2. Ion-implanted dual 64-bit dynamic shift register.

mentary MOS structures. Another unique feature of ion implantation that is directly applicable to read-only-memories is the possibility of using photoresist as a mask against the ion-implanted species.

5.3 ION-IMPLANTED READ-ONLY MEMORIES

Read-only memories are device structures that take advantage of the benefits of ion implantation to improve circuit performance. A 2048-bit, fully decoded, read-only memory built by ion-implantation techniques is shown in Figure 5.3. This device is arranged as a 512-word, 4-bit read-only memory with input and outputs compatible with TTL devices. The semiconductorchip measures approximately 110 $\times$ 140 mils.

Two external clocks are used. The address inputs must be valid after input Φ_1 ends. Before Φ_2 begins, the complements are formed on the chip and the decoding is completed. When Φ_2 becomes true, the selected drain and gate lines are charged. If a memory transistor is persent (a "1"), it will cause the memory to discharge after Φ_2. The TTL-compatible output stages are flip-flops that will change state only during Φ_1 if necessary. Thus the outputs are dc levels and not precharged, making system interconnection and operation less constraining. Total power dissipation is less than 400 mW at high-speed operating conditions.

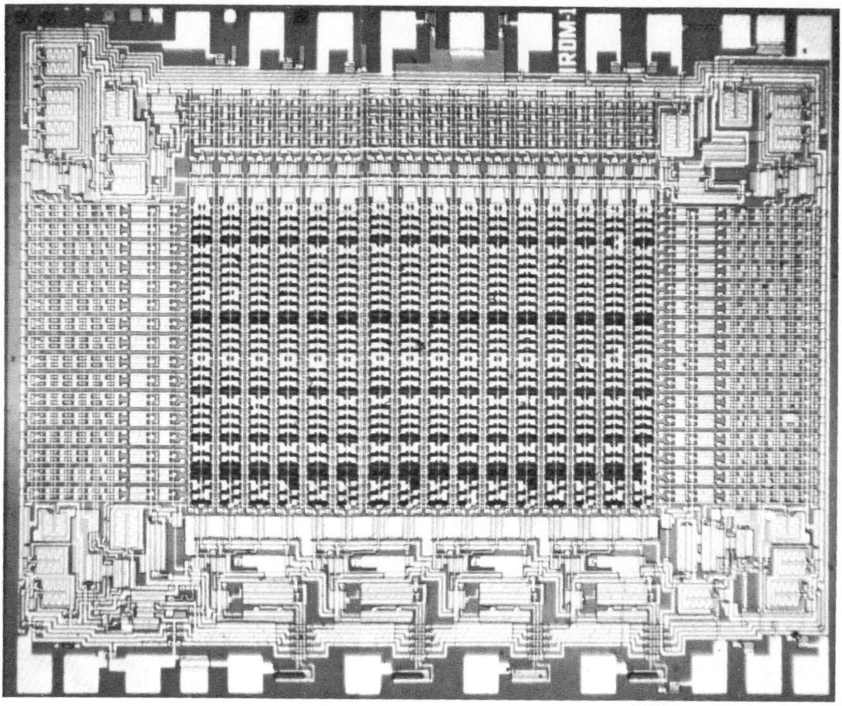

Figure 5.3. Fully decoded 2048-bit ion-implanted read-only memory.

In addition to lower capacitances on the individual devices, a unique feature of ion implantation is the word-encoding techniques used in fabrication. The 2048 bits of memory are stored in an area described by the center portion of the chip photograph. The periphery contains decoding logic and interface circuitry. The binary words to be written into the read-only memory are formed after wafers have completed metalization. In fact, wafers that have completed metalization and are ready for encoding may be stored indefinitely and encoded upon request.

Encoding consists of applying photoresist over the completed wafers and exposing through a photomask so that the unexposed photoresist can be removed over the areas where bits of the memory are to be ion implanted. Ion implantation then completes the MOS devices and determines the permanent logical "1" states. Subjecting the photoresist-coated, exposed, developed, and etched wafers to ion implantation causes areas from which the photoresist has been removed to implant in a normal fashion. After complete photoresist removal and annealing have been accomplished, the read-only memory is completed and ready for wafer testing.

With standard *p*-channel technology, the preferred method for determining memory logic states is to redefine gate areas with the gate etching mask.

Thus wafers must go through the critical gate etching-and-oxidation step, the contact-hole etching, and the metalization processes. A second standard technique for forming the logic states of the read-only memory requires only photomasking of metalized wafers and etching to delete gate metal from those devices to be disconnected. Because of possible surface contamination in exposed thin-oxide areas, the second technique is generally less reliable than etching the gate oxide. Therefore, ion implantation through a photoresist mask on completed wafers offers a third and possibly the best technique for completing the desired bit pattern.

5.4 CONCLUSIONS

Ion implantation applied to *p*-channel MOS read-only memories may provide the ultimate in gate self-registration with inherent improvements in size, speed, and power for the array. Furthermore, ion implantation promises to be the ideal semiconductor technique for encoding the permanent bit pattern required of the final ROM array. Future developments with this technology will allow implementation of an *n*-type dopant and then the possible exploitation of *n*-channel arrays with their increased speed–performance characteristics. The possibility looms for implanted complementary *p*- and *n*-channel devices to allow very low-power, high-speed, dense MOS arrays. Future performance improvements will be realized as designs are optimized for the basic implantation processes and as tighter dimensional control becomes generally available on a production basis.

REFERENCES

[1] Bower, R. W., Dill, H. G., Aubuchon, K. G., and Thompson, S. A., *MOS Field Effect Transistors Formed by Gate Masked Ion Implantation,* Hughes Research Laboratories Note, Newport Beach, Calif.

[2] Bower, R. W., *Applications of Ion Implantation Doping to the Planar MOS Integrated Circuit Technology,* Hughes Research Laboratories Note, Newport Beach, Calif.

[3] Moyer, N. E., Bower, R. W., and Dill, H. G., *High Speed MOS Integrated Circuits Utilizing Ion Implantation,* Hughes Research Laboratories Note, Newport Beach, Calif.

[4] Toombs, T. N., Dill, H. G., and Hartman, J. M., *A New 10-MHz Ion Implanted MOS Read-Only Memory,* Hughes Aircraft Company Note, Newport Beach, Calif.

[5] "Hughes Implants Silicon–Carbide Diode at Room Temperature," *Electronics Magazine,* March 30, 1970, p. 33.

6

MOS Read-Only Memories

by Michael R. McCoy

Electronic Arrays, Inc.

The first MOS read-only memory available was small (256 bits) and slow (2-μs access time), and was scarcely capable of performing any function that could not be done better, faster, and cheaper in some other way. But progress has been steady and MOS read-only memories can now be purchased in configurations providing speeds up to 750 ns, with capacities as large as 4096 bits, and at a cost that cannot be approached by any other technology. Such read-only memories are being delivered in volume. Some are easily programmable to any fixed storage pattern that the user desires and this type of memory will be described later in this chapter.

6.1 WHAT IS A READ-ONLY MEMORY?

A read-only memory is a circuit that can take a digital code at its input terminals and provide a unique digital code on its output terminals. Usually the relation between the input and output codes is fixed and can be changed only by replacing all or part of the circuitry. The only difference between a read-only memory and a read/write memory is the level of difficulty involved in changing the stored information.

Most read-only memories are also random-access memories, which means simply that all stored information is accessible in a fixed amount of time. It has been common practice (at least in the United States) to refer to random-access read/write memories simply as random-access memories.

That this practice is an oversimplification will become obvious with the increasing use of the random-access read-only memory.

Read-only memories are not new, although their usage in the past has been rather limited. The most common form of read-only memory is the diode matrix, which has been used since the early days of diode production. More recently the transformer type of read-only memory has found its way into microprogrammed digital processors. MOS read-only memories will replace some of the present forms in those applications where its cost and size advantages are significant; more often, however, the MOS read-only memory will create new applications that were not feasible prior to its development.

6.2 THE MOS READ-ONLY MEMORY

MOS technology is ideal for the fabrication of read-only memories. The high density attained in the memory matrix cannot be achieved by any other present technology. In addition, the circuitry required for address decoding and input–output buffering can be included on the same substrate containing the memory matrix. The result is a very dense, low-cost array that can be easily used. A block diagram for an MOS read-only memory is shown in Figure 6.1. The memory illustrated is organized as a 16-word by 2-bit/word read-only memory. Such a configuration is too small for economical production, but its simplified structure makes it easier to explain circuit operation.

The basic storage element of the MOS read-only memory is the MOS transistor. The memory matrix of Figure 6.1 has 32 possible transistor locations; hence it can be coded with 32 bits of information. To code a logical "1" in a particular cell, the designer specifies that a transistor be located in that position. To specify a logical "0" in a cell, no transistor is inserted. Insertion or deletion of a transistor in the matrix can be accomplished by changing a single photomask that is used in the MOS fabrication process.

Because it is desirable to keep the number of circuit inputs to a minimum, almost all available MOS read-only memories have full address decoding on the same substrate as the memory matrix. Note that in Figure 6.1 a coincidence-select technique is used to address a particular cell. This technique is used for two reasons: the length–width ratio of the memory matrix is more manageable, and three two-level decoder circuits require less circuitry and therefore are less costly than a single four-level decoder.

In the illustration of Figure 6.1, the Y select circuits fully decode the two Y addresses (Y_{A1} and Y_{A2}) into four Y cell-select lines 1, 2, 3, 4. There are two identical Y select circuits: one for each bit of the 2-bit word. The single X decoder provides a select signal on one of its four outputs.

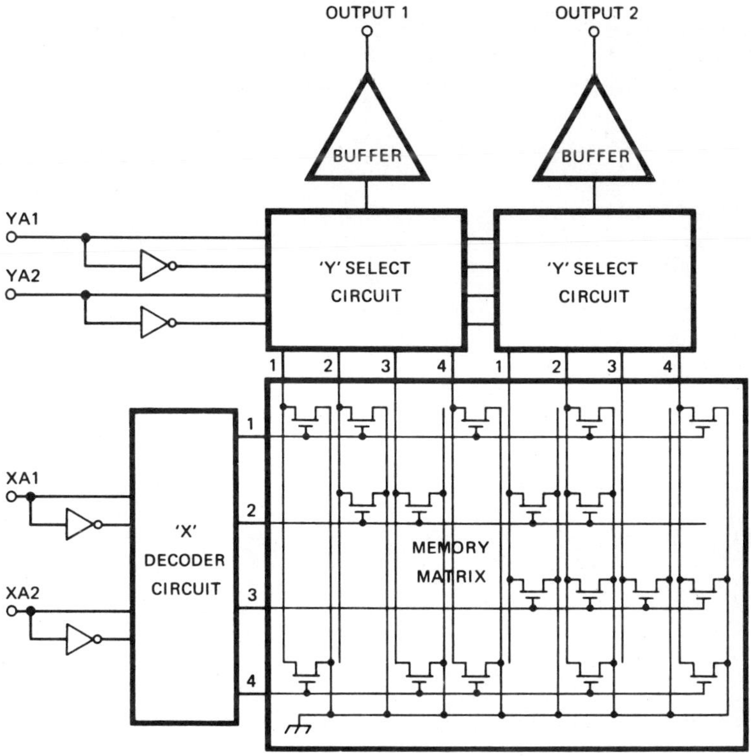

Figure 6.1. Block diagram of a 16-word, 2-bit/per word MOS read-only memory. A logical "1" in a cell is represented by a transistor at the intersection of the decoded X and Y address lines.

Readout of a particular cell in the array is accomplished by sensing for current flow in the cell at the intersection of the selected X and Y lines. In the illustration of Figure 6.1, two cells are always accessed simultaneously because of the 2-bit word configuration.

Two different circuit techniques can be used in the design of the decoder and buffer circuits—dc (unclocked, asynchronous) and dynamic (clocked, synchronous). Although sometimes the dc circuits are easier to use because of the lack of a requirement for clock pulses to activate the array, the dynamic circuits are smaller, which results in lower cost, faster access, and synchronous operation. Synchronicity allows use in sequential applications such as counters and sequence generators where a dc circuit cannot be used without the addition of external storage registers.

A photomicrograph of a 2560-bit read-only memory is shown in Figure 6.2. A section of the memory matrix is expanded so that a bit cell can be

Figure 6.2. Photomicrograph of the EA3501, a 512 by 5 MOS read-only memory, which is coded as an ASCII to a 5 × 7 dot matrix character generator. A portion of the memory matrix is enlarged to show the memory cells in more detail. The size of the array is 1.6 × 2.4 mm.

clearly identified. The signal lines of the matrix provide all the essential components for the MOS transistors that they interconnect. The decoded *X* address line serves as the gate electrode, the decoded *Y* select line functions as the diffused drain electrode, and the diffused ground return is the source electrode.

A photomicrograph of a 4096-bit MOS read-only memory is shown in

Figure 6.3. In mid-1970 this represented the largest single-chip memory in production. The triangular pattern in the memory matrix is a test configuration used to characterize for worst-case access time. The functional nature of the various sections of the array is shown in Figure 6.4.

6.3 THE CUSTOM-BIT PATTERN

One of the processes used to produce custom-bit-pattern read-only memories is shown in Figure 6.5. Typically, the first step in the process is

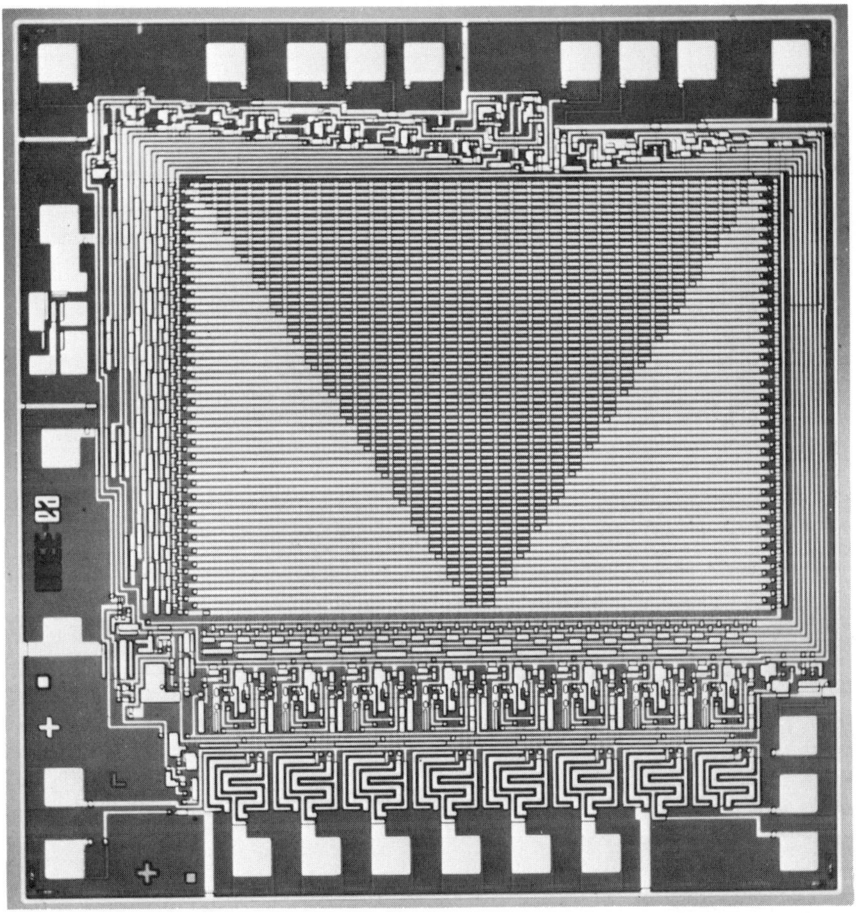

Figure 6.3. Photomicrograph of the EA3300, a 4096-bit MOS read-only memory in a 512 × 8 configuration. The triangular pattern in the memory matrix results from the use of a memory pattern to characterize worst-case access time. The die measures 2.2 × 2.4 mm.

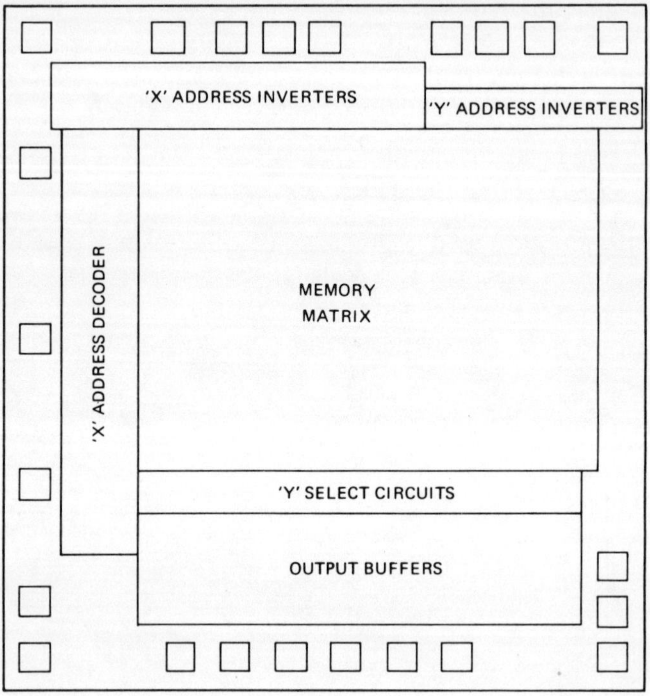

Figure 6.4. Block diagram of the 4096-bit EA3300 showing the location of the various components of the array. Refer to Figure 6.3 for comparison with a photomicrograph of the array.

completion of the customer coding sheet. This coding sheet specifies the input–output arrangement required of the custom read-only memory. The coding sheet is transferred to a computer, which provides the three separate outputs shown. A relisting of the input data is returned to the customer for verification. This verification step ensures that the data the computer used to generate the plot and test tapes are correct. The plot output is a scale drawing created by an automatic plotter and showing the location of each transistor to be inserted in the memory matrix. The test tape, also computer generated, consists of a punched paper tape that is used to directly load automated test equipment with the custom read-only memory outputs in order to perform a full functional test.

The next step in the process is fabrication of a custom mask. The plot output is used in conjunction with a master, typically 400*X* artwork, to create an image of the photographic mask with the customized memory matrix. This artwork is then photoreduced and stepped into an array of images for use in wafer fabrication. Silicon wafers, containing an array of read-only

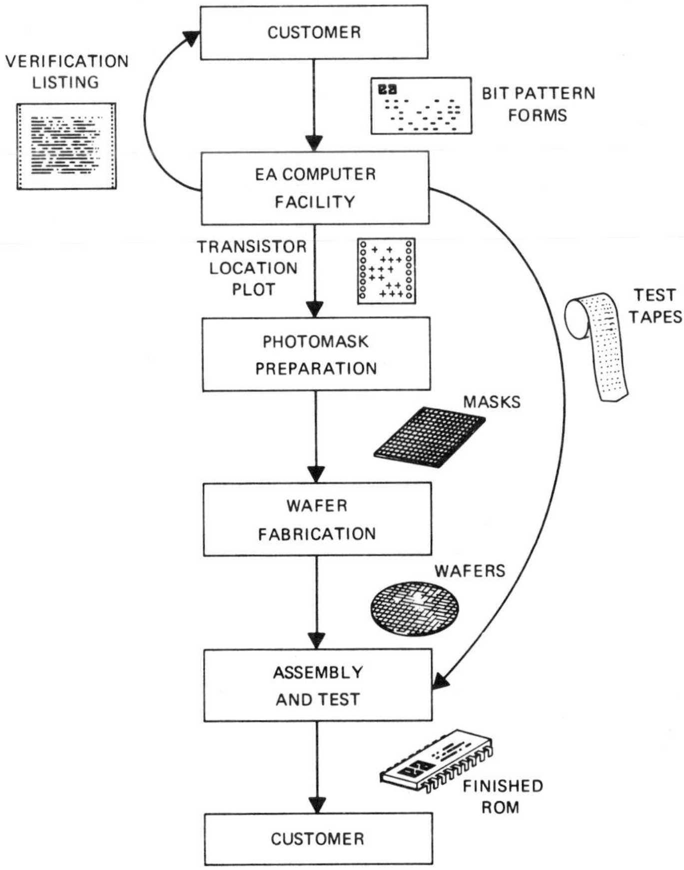

Figure 6.5. The custom-bit-pattern ROM fabrication process. Time required from start to finish is six to eight weeks.

memories, are then fabricated in the normal manner using the custom mask.

In order to avoid packaging circuits that are nonfunctional, each read-only memory on the wafer should be tested prior to packaging, using the test tapes produced at the computer center. Nonfunctional or marginal circuits can be marked and eliminated before packaging. The tested dice are then packaged, tested again, and forwarded to the customer.

The whole process requires between six to eight weeks. Every effort should be taken to reduce the probability of an error. The customer can be given an opportunity to check an output of the computer generated at the same time that the plotter output and test tapes to be used in fabrication are generated. Provided the customer does a careful job of verification, the only reasonable possibility of error exists in the mask fabrication process. Even if an error were to occur in the preparation of masks, the error should be caught at an intermediate stage when the wafers are first tested, thus keeping the loss of time to a minimum.

6.4 APPLICATIONS FOR THE MOS READ-ONLY MEMORY

The applications of MOS read-only memories are numerous, and new uses are being discovered at an accelerating rate. Almost all new character generators are being designed using MOS read-only memories to convert the digital character code to a screen code for dot-matrix characters. Another exciting application is in the fixed storage required for a microprogrammed digital processor. All forms of code converters are prime candidates for implementation as read-only memories.

The use of the MOS read-only memory in the character generator of an alpha-numeric display system is shown in Figure 6.6. Here, the read-only memory converts the digitally coded alpha-numeric representations in the data store into a dot-matrix code that visually represents the characters. For simplification in describing the function of the system, let us assume that all that is going to be displayed on the face of the cathode ray tube (CRT) is the three-character set shown.

During the first sweep (line 0), three lines from the synch-and-timing generator indicate to the read-only memory that line 0 is being displayed. As long as the line 0 code remains at the read-only memory inputs, no dot pattern will be displayed, since line 0 corresponds to the vertical space between characters. After the line 0 sweep is completed, the line 1 sweep is started. The first character code (A) is clocked into the ROM from the refresh memory and in conjunction with the line-1 code from the synchronization-and-timing generator causes a 00100 to appear on the read-only memory outputs. As the sweep generator causes the electron beam to sweep across the CRT face, the 00100 code from the read-only memory is shifted into the CRT blanking circuit, causing a single dot to occur on the CRT face in the middle of the line-1 position of the first character field.

Next, a new character (B) is shifted into the character code inputs of the read-only memory, causing a 01111 code to appear on the ROM outputs. The code is then shifted into the blanking circuitry, causing four dots in the first four-dot locations of line 1 in the second-character field. After the line-1 dot code for the third character is written, horizontal sweep return is initiated to start line 2.

The same three character codes will be sequenced into the read-only memory, but the new line code will cause dot representations appropriate for line 2 of the character field. This process is continued through line 7, when vertical sweep return occurs and starts another cycle.

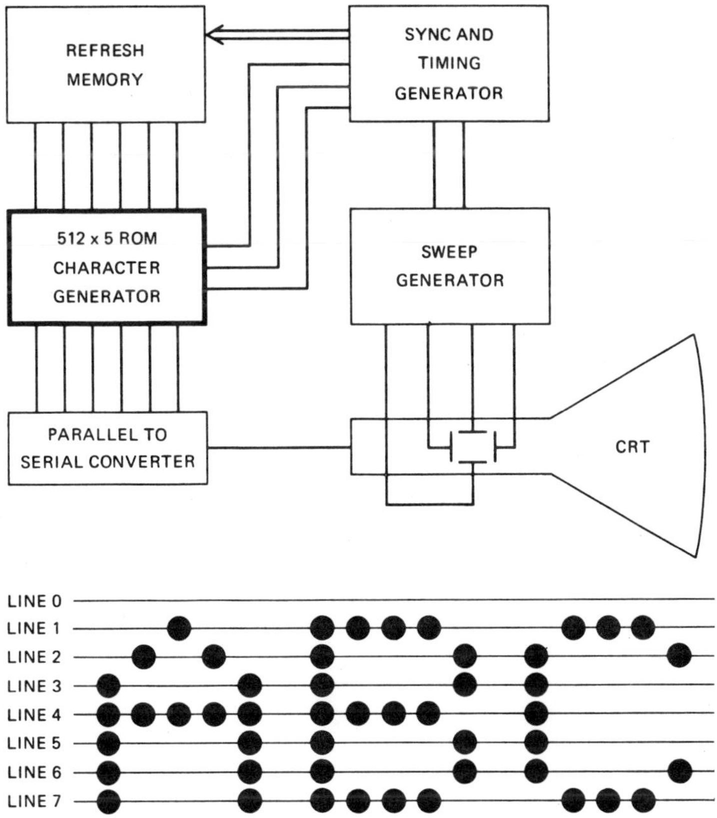

Figure 6.6. Block diagram and example of the output pattern of a dot-matrix character generator. The single ROM chip shown in Figure 6.2 is designed to perform the function of this ROM.

The standard, off-the-shelf, 512-word, 5-bit read-only memory shown in Figure 6.2 contains a memory pattern that converts a 6-bit ASCII code and a 3-bit line code into sixty-four 5×7 dot-matrix characters of the form shown in Figure 6.6. Changes in either the input code or the output code can be implemented with a custom bit pattern. The IC in Figure 6.6 is the EA3501, which is capable of operating at 1.3 MHz. To operate at higher rates, two 3501 units can be multiplexed. More complex dot patterns and more characters can be added by using more than one read-only memory.

A simplified block diagram of a microprogrammed digital processor is shown in Figure 6.7. The processor contains two read-only memories. One is the sequence ROM, which is essentially a sophisticated counter; and the other is the control ROM, which converts the sequence state into control commands for the arithmetic unit. In general, the processor operation is as

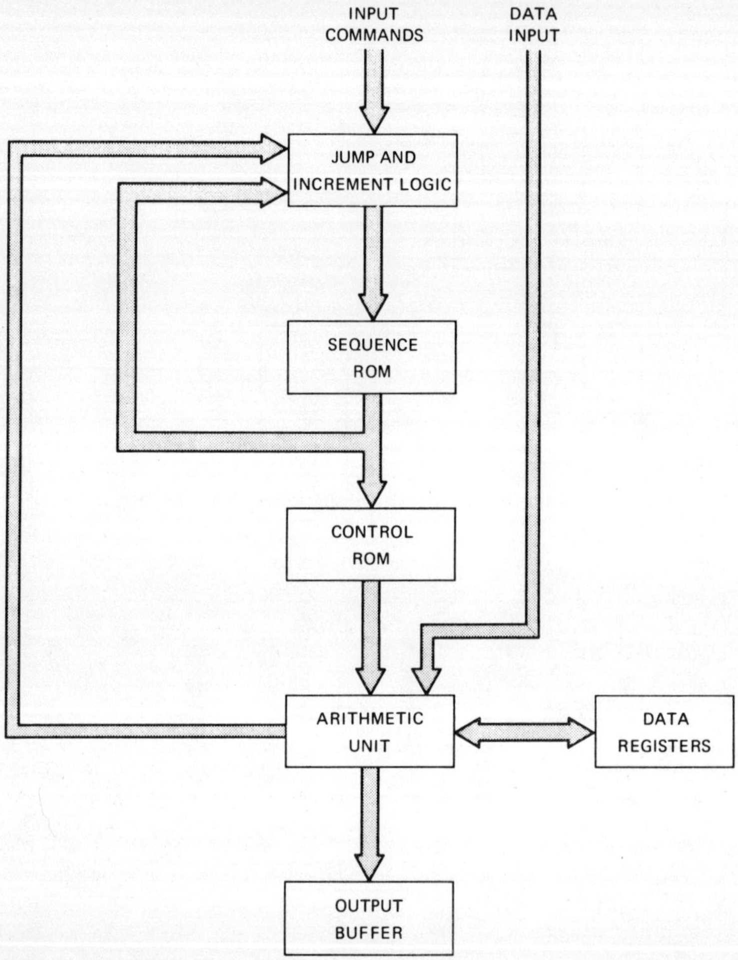

Figure 6.7. Functional flow diagram of a microprogrammed digital processor. Microprogrammed digital processors are more flexible and often much less expensive to produce than an equivalent "hard logic" system.

follows. First, an input command forces the sequence ROM into a particular initial state. The control ROM interprets the state and causes the arithmetic unit to perform an operation. The jump-and-increment logic detects when the sequence state is to be changed and initiates such action. The new sequence state is again interpreted by the control ROM and new commands are given to the arithmetic unit.

A processor of this type has many advantages. The arithmetic unit is simplified since the number of basic operations it must perform are limited.

Having sequence and control logic implemented in read-only memories offers great flexibility. New operations can be added and design changes made simply by changing the memory pattern, leaving the mechanical construction of the system completely intact. The cost of the system is reduced since the sequence and control read-only memories replace a great deal of logic. Many different functions can be performed by the same basic processor. The data registers and output bufler can be changed to fit the application, with the basic processor left intact.

Another major application area for read-only memories involves code converters and function generators. Many major manufacturers of digital equipment use their own character codes. A read-only memory can easily be used to convert one code to another. Two common character codes in use in the United States today are the American Standard (ASCII) and the code used in the IBM Selective typewriter. ICs are available, such as the EA3101, which convert ASCII code to Selectric code or Selectric to ASCII, depending on the state of one of its address inputs. A sine Φ generator is easily implemented with a read-only memory. An 8-bit sine generator could be implemented with a single 2048-bit read-only memory. An 8-bit representation of Φ could be applied to the inputs, with a resulting 8-bit representation of sine Φ occurring on the device outputs.

6.5 LOGIC AND COUNTING FUNCTIONS

Read-only memories can perform logic and counting functions as well. Any logic circuit with n inputs and m outputs can be replaced by a read-only memory with n inputs and m outputs. Any counter sequencer with n inputs and m flip-flops can be replaced by a sequential ROM with $n + m$ inputs and m outputs. Since any digital system can be constructed as logic and counters/sequencers, it is easily seen that any digital system can be completely constructed of read-only memories. The question thus becomes not one of whether a read-only memory can be used, but whether it is the most economical approach and whether is can best meet the performance objectives.

6.6 JUDGING MOS READ-ONLY MEMORY PERFORMANCE

Selection of a technology usually involves a weighted analysis of the following factors: (1) cost (present and projected), (2) performance, (3) physical size, and (4) reliability.

The approximate costs of a 4096-bit MOS read-only memory in 1970 and 1975 are depicted in Figure 6.8. Note the strong influence of quantity.

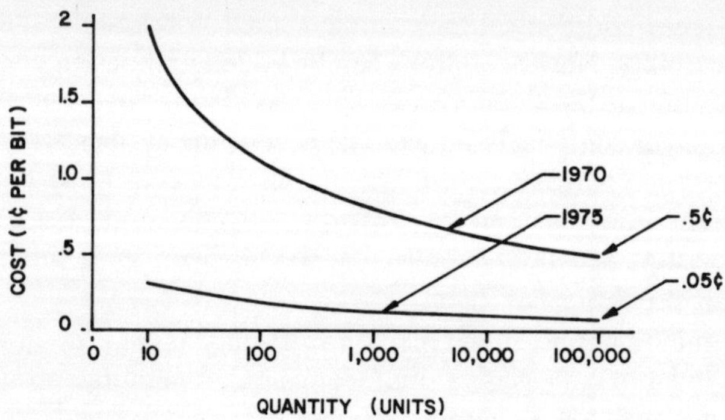

Figure 6.8. Present and projected cost per bit for a 4096-bit MOS read-only memory.

Circuits with lower bit capacities than this one will cost slightly more per bit.

The performance of MOS read-only memories is usually judged in terms of access time and power dissipation. The power dissipations of dc and dynamic-type read-only memories are far apart. A dynamic unit consumes on the order of 50 mW, whereas dc versions dissipate up to 500 mW. Current access time limits vary from 1.5 μs to 750 ns guaranteed over full voltage and temperature ranges.

6.7 THE FUTURE FOR MOS READ-ONLY MEMORIES

Crystal ball gazing is always a risky business, but the following should be achieved by 1975: (1) single-chip memories as large as 16,000 bits, (2) access times of less than 300 ns, (3) memory alterations accomplished in less than two weeks, (4) a wide variety of configurations to choose from, (5) high-density multichip packages providing assemblies of more than 64,000 bits, and (6) several hundred off-the-shelf MOS read-only memories precoded with commonly required memory patterns.

7

Applying MOS Memories

by David N. Leonard

Texas Instruments, Incorporated

The rapid evolution of MOS technology is forcing even the most skeptical to re-evaluate his ideas on the practicality of MOS memories. No one is predicting the immediate demise of magnetic memories, particularly in large main storages, but new concepts are emerging that take advantage of the performance and cost benefits that MOS has to offer. Data-manipulation systems using MOS random-access, read/write, scratch-pad memories are only the beginning.

The flurry of excitement in the development of up to 50,000-bit MOS random-access memory arrays (fabricated from complex MOS chips) has not distracted from other forms of MOS memories also receiving favorable attention. MOS read-only memories, for example, with thousands of bits per chip are already being used in such practical applications as trigonometric and logarithmic look-up tables for desk calculators and as character generators for data display systems.

With an order-of-magnitude speed increase imminent, the MOS read-only memory is also certain to enjoy widespread usage in computer microprograms.

A third (not yet popular) memory is the delay line. MOS shift registers, now available with hundreds of bits per chip, have revived interest in this memory form. Several shift registers running in parallel, with stored data being recirculated, provide the equivalent of a high-speed drum memory.

The use of MOS technology in content-addressable memories for associative data processing is now economically attractive, as well.

7.1 RANDOM ACCESS, READ/WRITE MEMORIES

An example of an MOS random-access read/write memory is the *p*-channel enhancement-mode memory chip show in Figure 7.1. This monolithic integrated circuit consists of 256 bits with only final-level decoding circuitry. The 16 × 16 matrix is organized as 256 words with 1 bit/per word. Like the sense winding on a core memory plane, one pair of digit lines goes to all cells on the chip. These two digit lines are used both for writing as well as for reading. Selection of a given bit for reading or writing is accomplished by addressing one of 16 *X* address lines and one of 16 *Y* address lines. Naturally, these 256-by-1 chips may easily be wired in parallel to form a memory array of any number of words or bits.

The basic storage cell in the 256-by-1 and most other MOS random-ac-

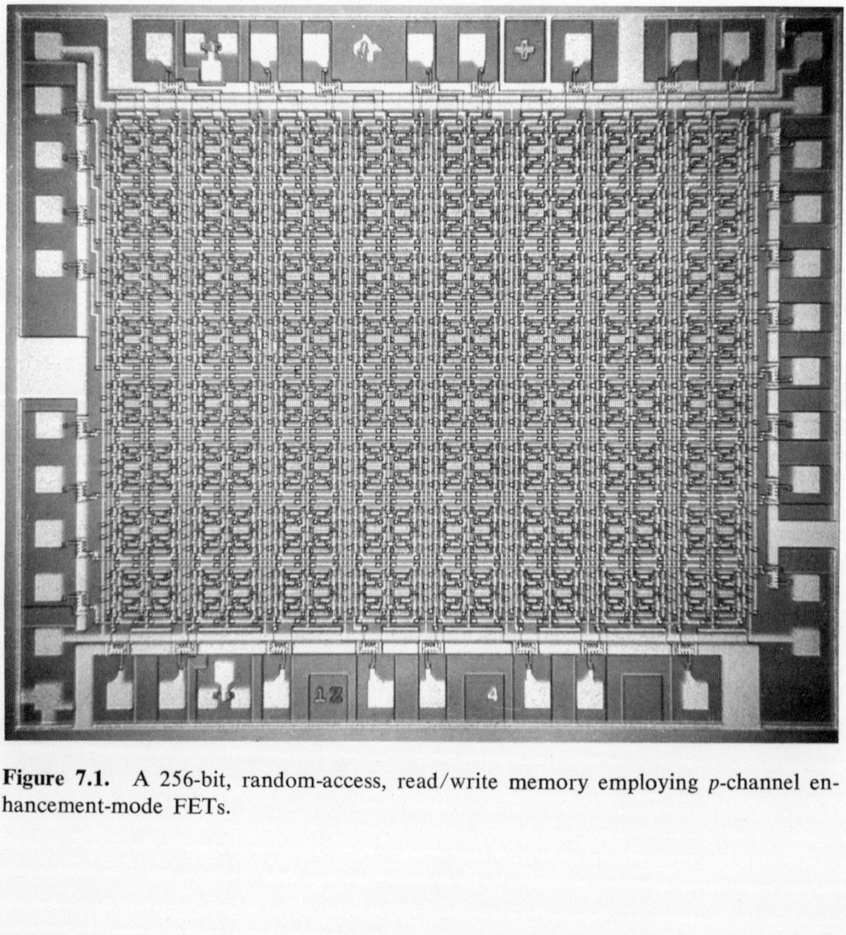

Figure 7.1. A 256-bit, random-access, read/write memory employing *p*-channel enhancement-mode FETs.

cess, read/write memories is a flip-flop (see Figure 7.2). In this figure the MOS substrate is returned to a positive voltage so that the normal role of MOS negative supply voltage and ground substrate is reversed for convenient interfacing to bipolar TTL systems.

Thus TTL system ground is effectively a negative voltage to the MOS chip. This is analogous to operating a bipolar *pnp* transistor with positive voltages, where the collector returns to ground and the emitter is positive. In this *pnp* analogy, the transistor will be "off" if the base is at the positive voltage and "on" if the base is brought to ground. For the MOS chip, the selection of a given cell is accomplished by grounding one of the *X* address lines and one of the *Y* address lines, causing transistors Q_1, Q_2, Q_3, and Q_4 in Figure 7.2 to turn on, thus connecting the flip-flop nodes of the desired cell to the digit lines.

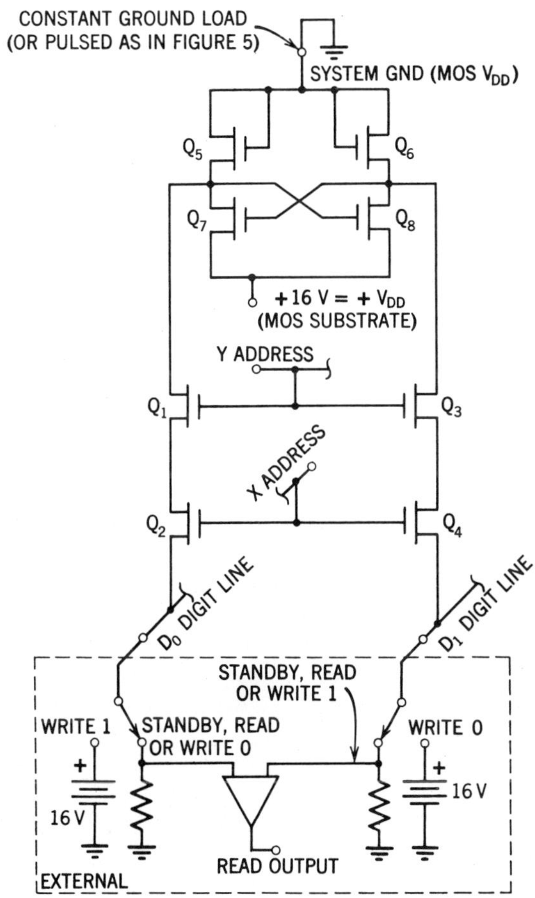

Figure 7.2. Basic MOS random-access storage cell.

Reading of the addressed cell is done by holding both digit lines through small-valued resistors to ground. Depending on the state of the flip-flop (either Q_7 or Q_8 will be on and the other will be off), at least 300 microamperes will flow in one or the other of the digit lines. Reading then is completed by sensing the differential current. Reading is nondestructive and actually tends to reinforce the existing state, since reading forces transistors Q_1, Q_2, Q_3, and Q_4 of the addressed cell to act as additional load resistors in parallel with the standby load resistors, Q_5 and Q_6.

Writing is performed by addressing a cell as if to read and holding one digit line at ground while bringing the other digit line to $+V_{SS}$. For the photograph in Figure 7.3, the cell at location ($X16$, $Y16$) was permanently selected. The middle trace is the D_1 digit line, the bottom trace is the D_0 digit line, and the top trace is the read output of a sense amplifier. This photograph demonstrates that the high-conductance addressing transistors allow the flip-flop to change state in less than 25 ns. The 25-ns digit pulses are alternatively changing the cell information with a write "0," a write "1," a write "0," etc.

The maximum speed of the MOS memory chip is limited by the propagation delay of the Y address voltage through a series of p-diffused tunnels. Figure 7.4 shows that the delay time to address the cell at the farthest end of a Y line is about 40 ns. Here, the top trace is the Y address voltage, and the bottom trace is the read output taken directly at the D_0 digit line rather than from the output of a sense amplifier. Recent process developments lower the resistance of the p-diffused tunnels and allow the 40-ns period to be reduced to less than 20 ns.

In order to conserve memory cell power dissipation, the MOS load

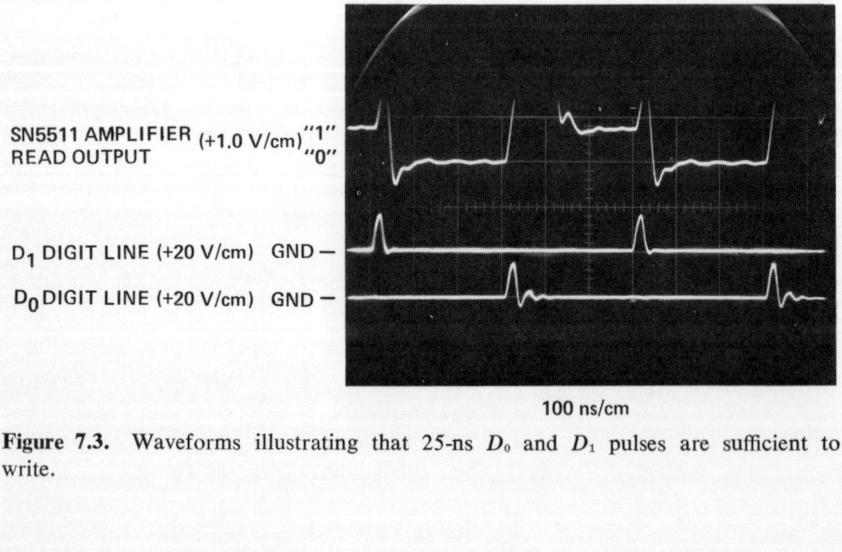

Figure 7.3. Waveforms illustrating that 25-ns D_0 and D_1 pulses are sufficient to write.

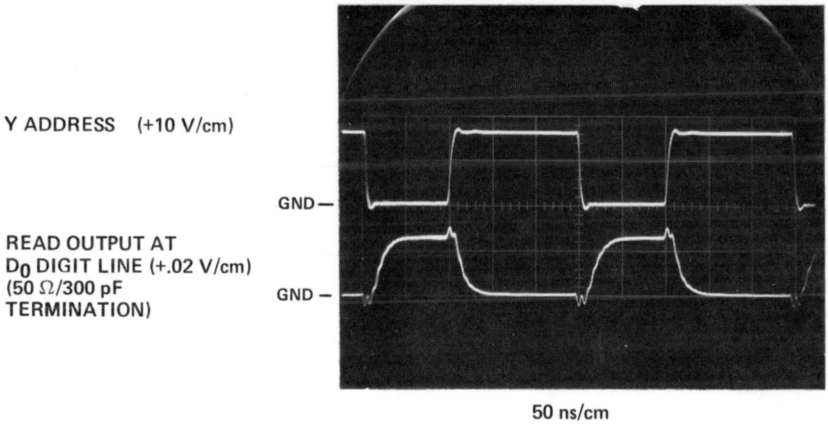

Figure 7.4. Waveforms illustrating that about 40 ns is required to address the most distant cell.

transistors can be pulsed with a 1% duty-cycle, 10-kHz pulse to ground, as illustrated in Figure 7.5, instead of the conventional constant ground load.

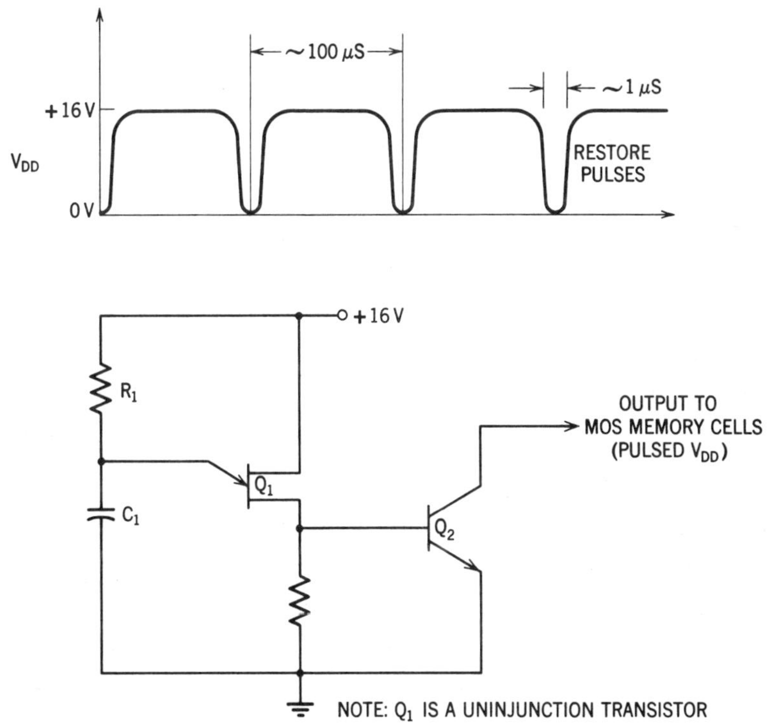

Figure 7.5. Power supply pulsing circuit and waveforms.

The pulses are referred to as restore pulses because they restore the charges on the flip-flop nodes. Th *npn* output transistor sinks a maximum of 10 mA per chip for about 1 μs to charge (restore) the memory cell matrix. Since cell addressing tends to charge the matrix rather than discharge it, readout can be performed asynchronously with the restore pulses. With such a pulsing circuit, the standby power dissipation will be in the vicinity of 6 μW per cell, which is 1.5 mW per chip. To avoid the large current surge that would occur if the restore pulse was delivered simultaneously to all the chips of a multichip memory, the memory can be divided into several sections that receive restore pulses in rotation.

7.2 WRITE/READ CIRCUITRY

The simplicity of the required external write/read circuitry is illustrated in Figure 7.6. The selected cell is written by supplying base drive to one transistor from an open-collector quad TTL gate. This action pulls the digit line to +16 V and changes the memory cell state. The other memory write line is held to ground by a 110-Ω termination. (Note: The particular TTL gate shown, SN7401, is not tested for an output voltage breakdown of 16 V. However, most of these devices exhibit breakdowns in excess of 20 V, making them usable in this application if tested beforehand.)

Line sensing is performed across the write-termination resistors R_1 and R_2, which are clamped with a diode to protect the amplifier input during writing and to enhance *RC* charging speed after writing. The sense amplifier (or comparator) determines the cell state by sensing which termination line has current flowing to ground.

7.3 APPLYING THE MOS CHIP TO A MEMORY SYSTEM

Suppose a memory organization of 1024 words with 52 bits per word is desired. With each MOS chip containing 256 words by 1 bit/word, four chips are needed for each bit plane. This group of four chips is organized as a 32 × 32 array. The four D_1 digit lines are wired together and the four D_0 digit lines are wired together, giving a pair of lines going to a write/read circuit for that bit plane. To achieve 52 bits/word, 52 such planes are required. The 32 X and 32 Y address lines are common to all planes, and a write/read circuit is required for each plane.

For a system of this size using bipolar TTL for the first and second levels of address decoding, the cycle time for reading or writing will be less than 200 ns.

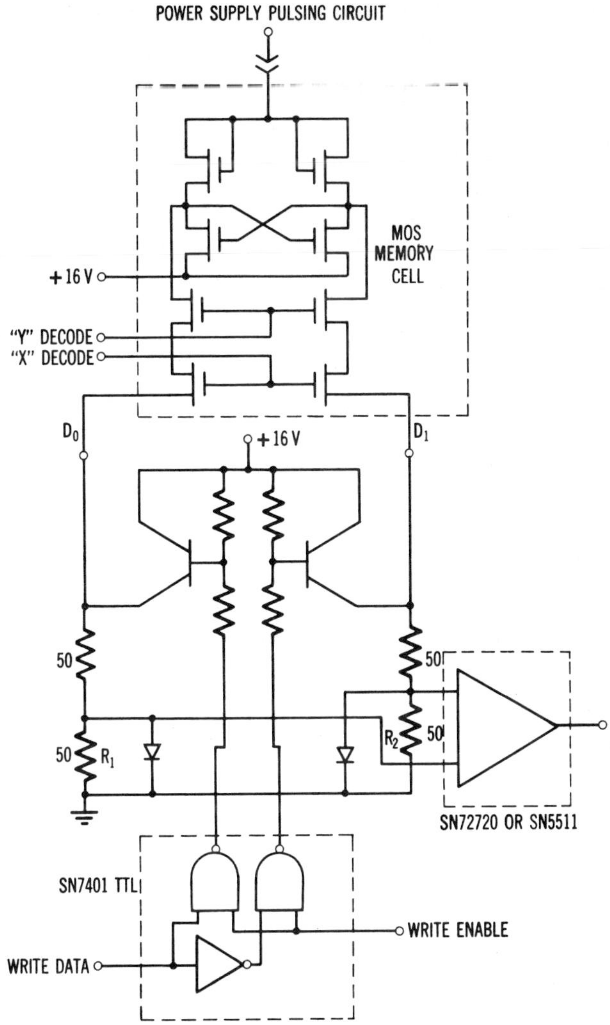

Figure 7.6. External write/read circuitry.

7.4 ADVANCED CIRCUIT ORGANIZATIONS

Most of the speed limitations of present MOS memories can be eliminated by a combination of process and circuit improvements. One of the limitations of the circuits in use today is the time delay of cell address information propagating down an *RC* transmission line in which resistance results from interconnection tunnels. This limitation can be removed by the use of double-level metalization.

A second practical limitation of some cells is the need to read from the same data lines used to write, since the transient from a write cycle makes high-speed reading difficult. This problem is eliminated by cell redesign in which the read line is separate from the write lines. With these two improvements, MOS memory cells are capable of very high speed performance, having a cycle time of less than 50 ns.

Double-level metalization will not only improve performance but also will permit the service lines to run on top of the storage cells rather than beside them, almost doubling the cell density. Combining this feature with smaller basic cells and with ever-improving photolithographic capabilities, chips having 1024 or 2048 bits are replacing the lower complexities in use today, reducing assembly and testing costs.

In those applications in which cycle times of several hundred nanoseconds are tolerable, costs are reduced by having all the address decoding and sense circuitry on the MOS chip.

An attractive technique for assembling highly complex MOS memories into a compact, stable package involves the use of beam leads and multilevel ceramic substrates. With this technique it should become economical to assemble an entire memory, or blocks of 50,000 to 100,000 bits, on a single ceramic substrate of perhaps 4 × 4 in.

7.5 STATIC READ-ONLY MEMORIES

Static read-only memories can be economically fabricated with thousands of bits per chip. This achievement is possible because, when writing is not necessary, the storage of a "1" or a "0" only requires one transistor. (Static refers to the capability of dc operation.) This dc operating capability eliminates the need for the special clocking required by dynamic MOS circuitry.

Useful as look-up tables and microprograms and as character generators, ROMs can also be put to work as random logic. A read-only memory with thousands of bits of storage can be implemented with a diode matrix or with bipolar logic, but in such cases the memory is large and expensive. Magnetic-core read-only memories require special drive and sense circuitry and are difficult to modify for different customer requirements. MOS ROMs, however, achieve this customization with a single mask modification and provide significant savings in cost, size, weight, and power.

Figure 7.7 shows the gate mask for a device that is typical of available static MOS read-only memories. A given input code (address) produces a preprogrammed output word. This device was specifically designed for use in a wide variety of display systems, from billboard displays to high-speed CRT data terminals.

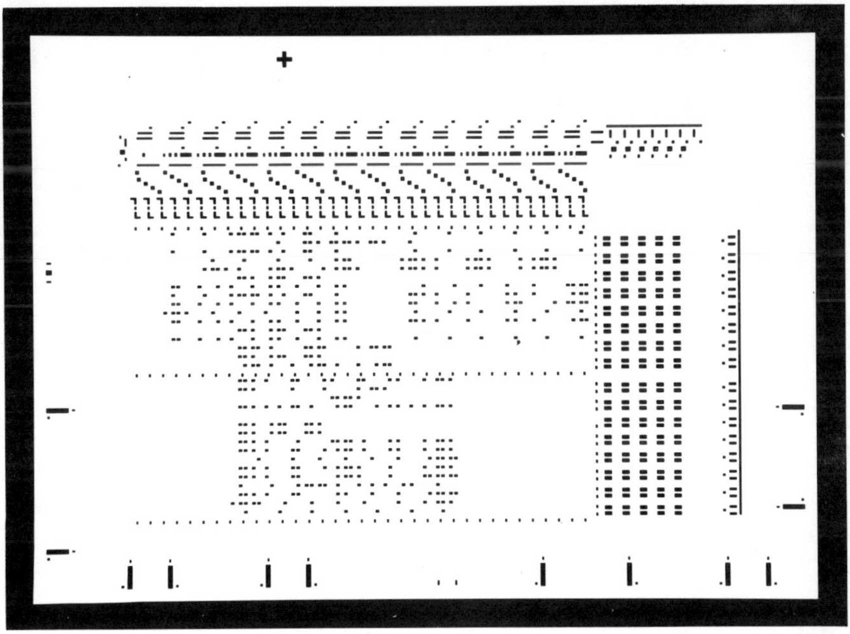

Figure 7.7. Gate mask pattern for a 2240-bit MOS read-only memory.

When a ROM is used in a display system, the output word is used to describe an alpha-numeric character. Consider, for example, the 5 × 7 dot matrix of Figure 7.8, which is presenting the letter *H*. If a lamp exists at each of the 35 dot positions, then lamps forming the letter *H* are on while the others are off. The on lamps correspond to "1"s of a 35-bit output word, whereas the off lamps correspond to the "0"s of the word. Having all 35 output bits appear simultaneously is useful for light matrices and print-head drivers. To generate a full keyboard set of 64 characters requires a memory with 2240 (64 × 35) bits of storage. Such a device (actually

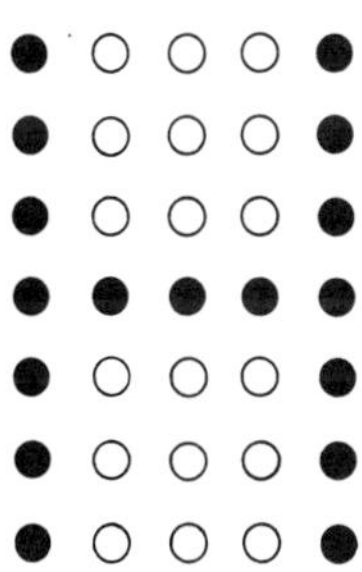

Figure 7.8. A 5 × 7 matrix displaying the letter *H*.

having 76-character capability) is available. However, in most display applications, having all 35 outputs appear in parallel is neither necessary nor convenient. For instance in a TV raster-scan display, the output should consist of five parallel outputs chosen on a row-selectable basis, where seven sweeps are required to display a 5×7 character. Since most TV displays use interlaced scanning, the row selection must alternate with rows 1, 3, 5, and 7 in one field and 2, 4, and 6 in the other. For moving billboards such as stock market displays and for *X–Y* CRT displays, a column-select rather than row-select capability is desired. With either the row-select or the column-select feature, the ROM still requires 2240 bits of storage to handle 64 characters.

The ROM of Figure 7.7 provides column-select, that is, the desired column is made available at the output by driving one of five column-select inputs. The memory can be internally organized to provide sixty-four 5×7 characters or thirty-two 5×14 characters. Total power dissipation is under 400 mW, and an entire character can be extracted in less than 1.5 μs.

For applications requiring higher speeds, such as microprogramming, new circuit techniques have been devised that provide 50-ns access time for MOS chips containing more than 2000 bits of storage.

The almost astounding circuit complexity and performance of available MOS devices leads one to the question, where do we go from here? The answer is simple: to more complexity and better performance.

ACKNOWLEDGMENT

The author wishes to acknowledge contributions to this chapter by Bob Proebsting, Daniel Baudouin, Herschell Murry, and Gordon Hoffman.

8

MOS Content-Addressable Memories

by David N. Leonard

Texas Instruments, Incorporated

A constant-addressable memory (CAM) is a storage medium that can be accessed by searching for data content. To cite a simple example, consider the 5-word, 4-bit/word memory in Figure 8.1. Suppose the problem is to find all words that contain the data $10XX$ where X may be either a "0" or a "1"—that is, X is a "don't care" or masked interrogate bit position.

The conventional approach to solving this problem is to retrieve word 1 from memory, operate upon that piece of data to see if it matches the criteria, then retrieve word 2, operate upon that datum, etc., until all words have been searched. It is obvious that this conventional approach is quite time consuming if the memory contains many words. However, given a CAM with parallel search capability, then all words in the memory are simultaneously asked to perform a comparison ("match") with the argument register, but only in the bit positions corresponding to a "1" in the mask register. Each word has an interrogate output that indicates if the piece of data contained in that word matches the information that the memory is being interrogated for. The next operation could be to ask if any of the words match or to count the words that matched; or, the next step could be to read sequentially the words that matched to learn what data those words are holding in the masked bit positions.

The ability to search out or interrogate stored data on the basis of their contents can be a powerful asset in many applications. However, another

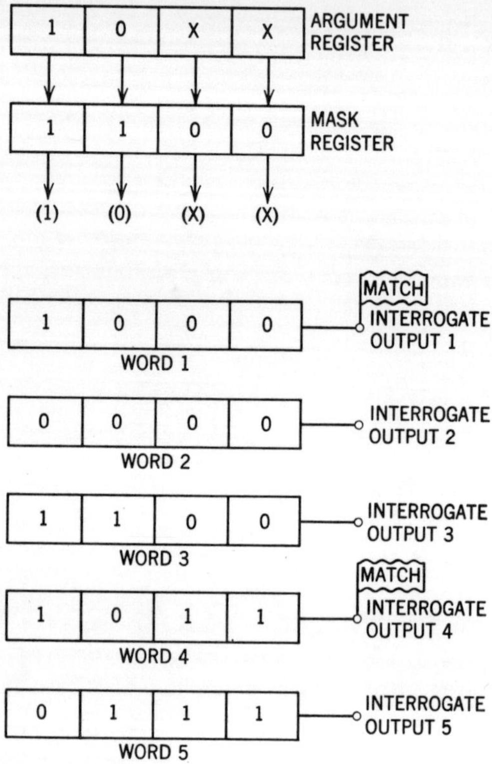

Figure 8.1. A single, 5-word per bit content-addressable memory.

facet often overlooked is that the CAM forms the basis for the more sophisticated case of truly general associative data processing (ADP). Associative data processing is defined as dealing with the interrelations among data and among sets of data. The simple match process (which a CAM has the ability to perform) can be iterated to perform more complex operations of set definition and set positioning, for example: greater than, less than or equal to, next greater, next lower, between the limits of, maximum, or minimum. Consequently, it is possible to rapidly extract from an ADP content-addressable memory not only those things in a subset that match a certain characteristic, but also those things nearest the desired attribute, those greater than, etc.

8.1 EXAMPLE

Consider, for example, an associative algorithm to extract the maximum-valued word from a 512-word, 24-bit/word CAM, in which the bit positions have been assigned binary weighting. The first operation could be to interrogate the most significant bit of all words for a "1" with all other bits defined as "don't cares" (masked interrogate); i.e., $1XXX\text{-}X$. If only one word indicated a match, then the maximum-word search would be completed. However, if several words indicated a match, then the CAM would be interrogated again, this time for $11XX\text{-}X$. The third possibility is that on the first interrogate no words indicated a match, in which case the second interrogate would have been for $01XX\text{-}X$. The net result is that, for a 24-bit/word memory, no more than 24 cycles is required to determine the maximum-valued word, regardless of the total number of words.

With a magnetic drum and some additional logic at the read heads, the CAM function could be performed sequentially, with each word of data examined one at a time. The disadvantage of the drum or of any type of sequential CAM is slowness. As seen in the above example, any complex set partitioning problem requires several searches of the entire memory. The obvious solution to this speed problem is to build a CAM that can be interrogated in parallel; but a parallel CAM requires logic at each bit position in memory. Such a memory was definitely not economical before the advent of MOS technology.

8.2 PRACTICAL MOS CAM

A practical MOS CAM organization of 8 words with 8 bits per word is shown in Figure 8.2. Packages may easily be wired in parallel to form a memory array of any number of words and bits. The memory is capable of three operational modes: write, read, and interrogate. The manner by which these modes are performed can best be understood by reference to the simplified 2×2 array of Figure 7.9 in conjunction with the following discussion.

Addressing of a word for reading or writing is a linear select scheme accomplished by applying a negative voltage to the appropriate word line, W. For each bit column there are two bit lines similar to the digit lines in the 256-bit random-access memory discussed in Chapter 7. For the CAM, these lines are used not only for reading and writing but also serve as interrogate inputs.

8.3 WRITE OPERATION

To write, a word is addressed and one of the other bit lines is pulsed to ground. Unlike the voltage convention adopted earlier in the discussion of the 256-bit random-access memory, the MOS substrate is now returned to system ground and all MOS power supply voltages are negative voltages.

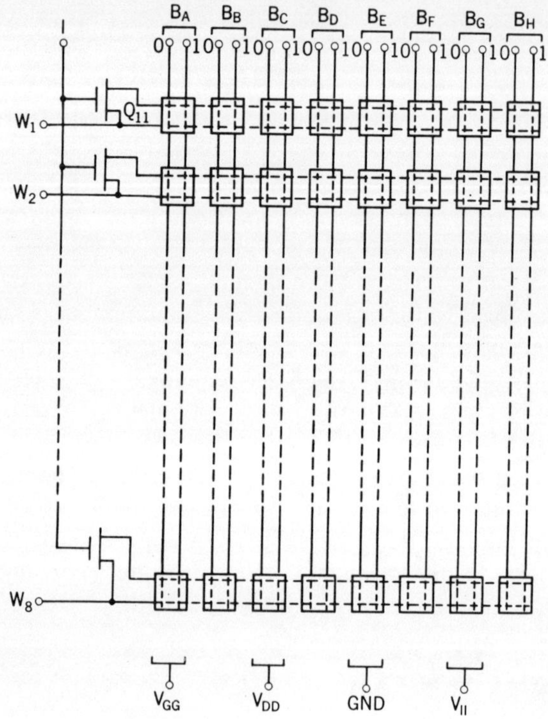

Figure 8.2. Content-addressable memory of 8-words, 8 bits/word.

An interesting feature that should not be overlooked is the capability of performing masked writing—writing new information into some bits of a selected word while leaving other bits unchanged. The information in a selected bit will be unchanged during a write cycle if both bit lines going to that bit are maintained at a negative voltage as if a read operation was being performed.

8.4 READ OPERATION

For an actual read operation, one word is addressed. The state of a cell at a given bit position is determined by holding each of the two bit lines at a negative voltage through a small resistance and sensing which one of the lines is conducting current.

For the interrogate mode, the entire memory is simultaneously searched by content to find which words match the word being searched for. The interrogate mode is flexible in that any bit columns can be defined as "don't cares." The *I* line shown in Figures 8.2 and 8.3 is an interrogate enable input. This line is grounded for read, write, and standby, but is taken to a negative voltage for an interrogate operation. To interrogate,

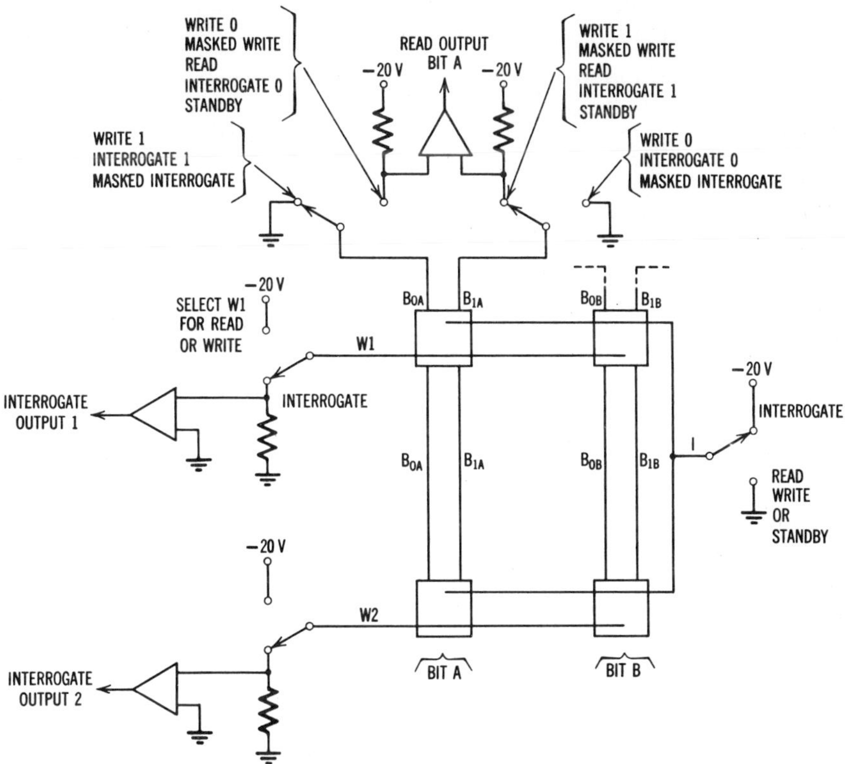

Figure 8.3. A CAM with two words, two bits per word.

each word line is terminated to ground through a small resistor, and appropriate voltages are applied to the bit lines. The input voltages required to operate the CAM are summarized in Table 8.1. If a word perfectly matches (equals) the interrogate information, then no current will flow through the W-line termination resistor. One or more mismatches will cause at least

TABLE 8.1 Definition of Input Voltage and Output Signals

Operation	Input Voltages				Output Signals
	I	W	B_0	B_1	
Write 0	0	$-V$	$-V$	0	
Write 1	0	$-V$	0	$-V$	
Masked write	0	$-V$	$-V$	$-V$	
Read 0	0	$-V$	$-V$	$-V$	Current in B_1 (300 μA min)
Read 1	0	$-V$	$-V$	$-V$	Current in B_0 (300 μA min)
Interrogate 0	$-V$	0	$-V$	0	Current in W indicates mismatch (100 μA min)
Interrogate 1	$-V$	0	0	$-V$	(same as above)
Masked interrogate	$-V$	0	0	0	No current in W from this bit
Standby 0	0	0	X	X	

100 μA to flow through the resistor, causing a voltage drop that is detected by the sense amplifier.

The basic CAM cell schematic is shown in Figure 8.4. All transistors in the 128-bit COM circuit of Figure 8.4 are *p*-channel enhancement-mode FETs. Transistors Q_1, Q_2, Q_9, and Q_{10} comprise a flip-flop for the storage of a "1" or a "0." Transistors Q_7 and Q_8 are selection transistors which, when turned on by application of a negative voltage to the *W* line, connect the flip-flop nodes of the memory cell to the B_0 and B_1 lines. When the *W* line is at O V, Q_7 and Q_8 are off, isolating the flip-flop nodes from the B_0 and B_1 lines. Transistors Q_3, Q_4, Q_5, Q_6, and Q_{11} perform the logic necessary for the interrogate function. For the interrogate mode, each *W* line is terminated through a small resistor to ground and Q_{11} is turned on. Transistors Q_3, Q_4, Q_5, and Q_6 compare the state of the memory cell flip-flop with the voltages externally being applied to the bit lines. By reviewing Table 8.1 in conjunction with Figure 8.4, it is clear how the CAM functions. Note that most lines are used both as inputs and as outputs. The word lines are inputs for write and read operations but are outputs for the interrogate operation. The bit lines are inputs for write and interrogate but serve as outputs for reading.

A 400-ns cycle time for a 128-word by 64-bit/word array has been achieved. An important footnote to this is that speed is almost independent of array size; the cycle time for a single 8 × 8 chip is slightly less than 400 ns, whereas the cycle time for a 512-word by 104-bit/word array is only slightly greater than 400 ns.

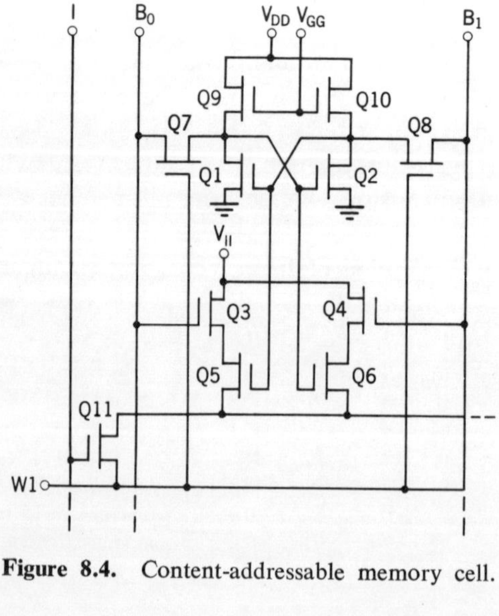

Figure 8.4. Content-addressable memory cell.

9

Complementary MOS ICs

by Donald V. DiMassimo

G-V Controls, Inc.

Let's review briefly a few basic characteristics of MOS insulated-gate field-effect transistors. These devices consist of two electrodes (source and drain) diffused into a silicon substrate. The source and drain are separated by a finite space, thus forming a majority-carrier conducting channel. A metal gate is placed above the channel and insulated from it.

An *n*-channel device is normally operated with the drain at supply potential V_{DD} and the source and substrate at ground, V_{SS}, potential. If a positive potential is applied to the gate, a field is created between the gate electrode and the substrate. Since the field must terminate on charges, electrons move through the substrate and accumulate in a layer immediately beneath the silicon surface. This layer is termed an inversion layer; i.e., the *p*-substrate is inverted to *n*. Thus an ohmic path is formed between the drain and source, and current flow is established between supply and ground. If, however, the gate potential is reduced to ground potential, the inversion layer is immediately removed and the device current is cut off. In this situation, the drain is heavily reverse biased, so that the only current flow is leakage, normally in the picoampere level.

In enhancement MOS devices, such as those used in complementary-symmetry MOS (COS MOS) arrays, the gate voltage does not have to be reduced to ground potential to cut off the device. Normally the cutoff point is established by the process above the minimum value required for good noise immunity. The gate voltage at cutoff is termed the threshold voltage, V_{th}.

A p-channel FET performs in a manner similar to the n-channel device except that all signs are reversed; i.e., the potentials are negative. The source and drain electrodes are p^+-diffusions in an n-type substrate; the induced layer of charges is termed an accumulation layer, and the carriers involved in conduction are holes.

9.1 DRAIN-CURRENT CHARACTERISTICS

The typical drain-current characteristics of an n-channel enhancement-type transistor are shown in Figure 9.1. Two current regions are defined: the triode region, where

$$V_{DS} < V_{GS} - V_{tn}, \tag{9.1a}$$

$$I_{DS} = K_n \, [2V_{DS} \, (V_{GS} - V_{tn}) - V_{DS}]^2 \tag{9.1b}$$

and the saturated region, where

$$V_{DS} \geqslant V_{GS} - V_{tn}, \tag{9.2a}$$

$$I_{DS} = K_n \, (V_{GS} - V_{tn})^2. \tag{9.2b}$$

In the above equations, the constant K_n is basically a function of design

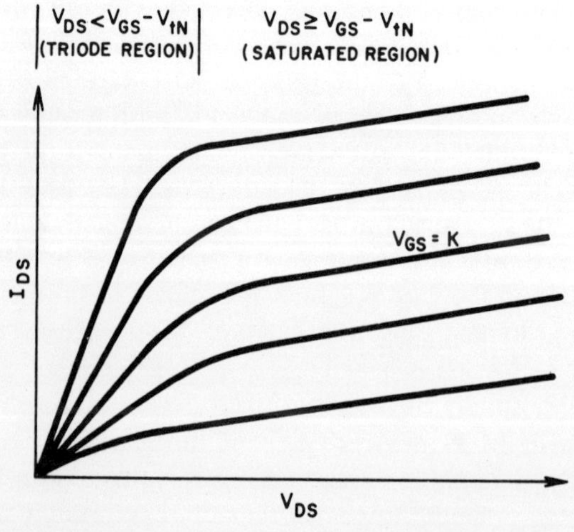

Figure 9.1. Typical drain current characteristics for an n-channel MOS device.

and processing and is given by $(\mu_n E_{ox} W)/2L\, T_{ox}$, μ_n = effective mobility of electrons, where E_{ox} = permittivity of insulating oxide layer, W = channel width, L = channel length, and T_{ox} = thickness of the insulating oxide (1000 Å typical).

The channel dimensions W and L are design variables. Substituting typical process variable values for μ_n, E_{ox} and T_{ox} results in

$$K_N = 5.7 \times 10^{-6} \left(\frac{W}{L}\right). \tag{9.3}$$

Typical dimensions for an n-channel device are $L = 0.25$ mil and $W = 3$ mils. If $V_{GS} = 10\ V$ and $V_{tn} = 1.5\ V$, then the saturation current from (9.2b) is:

$$\begin{aligned} I_{DS} &= 5.7 \times 10^{-3} \left(\frac{3}{.25}\right)(10 - 1.5)^2 \\ &= 4.9 \text{ mA.} \end{aligned}$$

The equations for p-channel devices are similar except for appropriate changes in sign and with V_{tp} and μ_p substituted for V_{tn} and μ_n. The slope of the drain-current characteristic in the FET triode region is termed R_{DS}. This value can be computed from:

$$R_{DS} = \left(\frac{\delta I}{\delta V_{DS}}\right)^{-1}_{V_{DS}=0} = \frac{1}{2K_N\,(V_{GS} - V_{tn})}. \tag{9.4}$$

For the n-channel device mentioned above, this value will be:

$$\begin{aligned} R_{DS} &= \frac{1}{(2)\,(5.7 \times 10^{-6}) \left(\frac{3}{0.25}\right)(10 - 1.5)} \\ &= 860\ \Omega. \end{aligned}$$

An approximate current–source equivalent circuit is shown in Figure 9.2.

9.2 THE COMPLEMENTARY INVERTER

The complementary inverter (Figure 9.3) will be used to illustrate the basic operation of COS MOS circuits. In this circuit, when V_{IN} is equal to V_{DD}, the n-channel device is turned on, the p-channel device is off, and the output is at ground potential. Conversely, when the input voltage is at

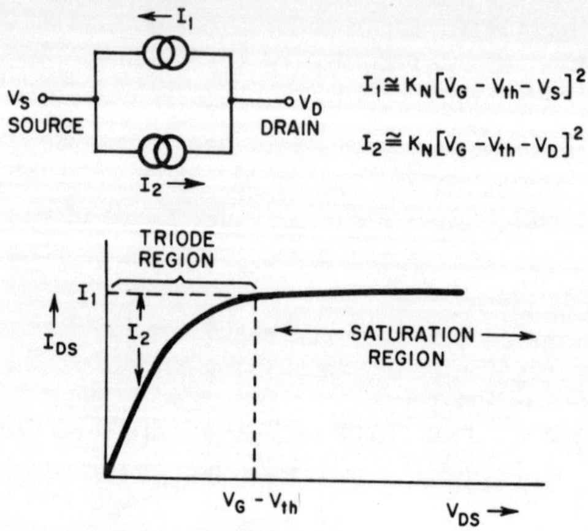

Figure 9.2. Current–source equivalent circuit for an *n*-channel device.

ground, the *n*-channel device is off, the *p*-channel device on, and the output is at V_{DD} potential. Note that only during the actual switching period is there a direct connection between V_{DD} and ground; thus under static conditions essentially no power is dissipated in the circuit.

The transfer characteristics for this device are shown in Figure 9.4. An important voltage when noise immunity is being considered is the voltage that, when applied to the inverter, will cause both devices to be turned on. This is termed the transition voltage (V_i in Figure 9.4). For optimum noise immunity, this voltage should be ½ V_{DD}. The transition voltage can be computed from the following equation:

$$V_i = \frac{\sqrt{\dfrac{K_p}{K_n}}(1 - |V_{tp}|) + V_{tn}}{1 + \sqrt{\dfrac{K_p}{K_n}}}. \tag{9.5}$$

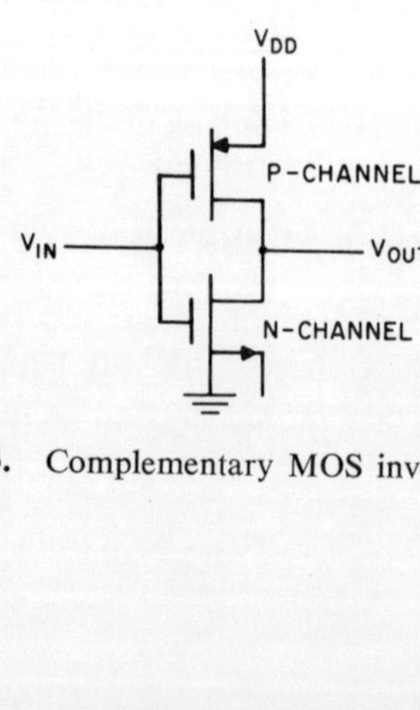

Figure 9.3. Complementary MOS inverter circuit.

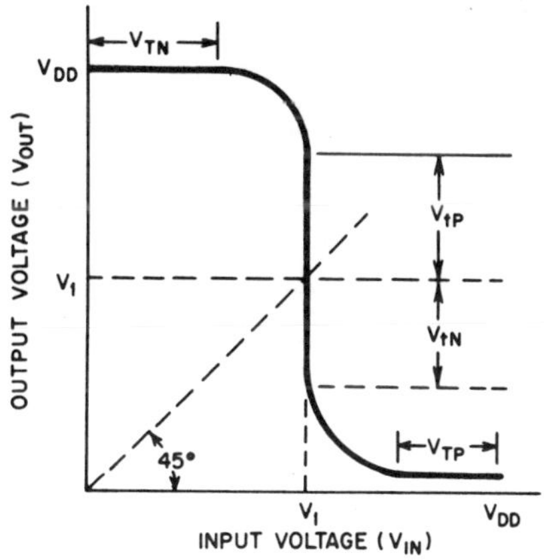

Figure 9.4. COS MOS inverter transfer characteristics.

Normally, in the design of a COS MOS array, the device channel length L is held to a minimum. This is done to maximize transconductance, speed, and packing density. The channel width W is then tailored to obtain the desired device characteristic. Therefore, substituting the average threshold voltages in (9.5) and inserting typical K_n, K_p constants, we have $V_i = \frac{1}{2} V_{DD}$ if $W_p \cong 2W_n$. The widths expressed here provide the width ratio desired for maximum speed.

The transition current I_i during the actual switching interval versus switching voltage characteristic is shown in Figure 9.5. The maximum current can be computed from (9.2):

$$I_i = I_p = I_n = K_n \, (V_{GS} - V_{tn})^2 \tag{9.6}$$

where $V_{GS} = V_i$.

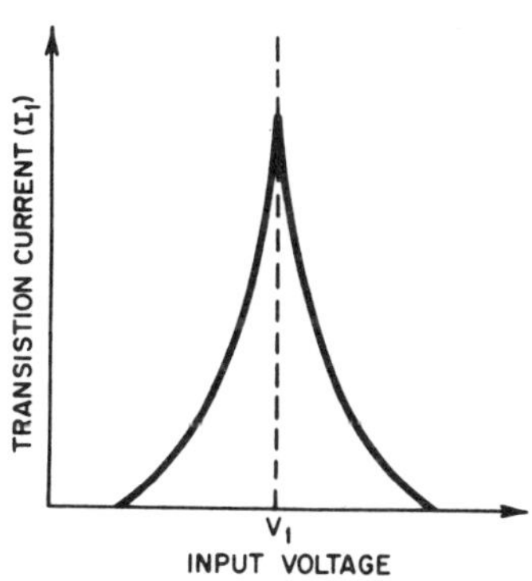

Figure 9.5. Transition current characteristics.

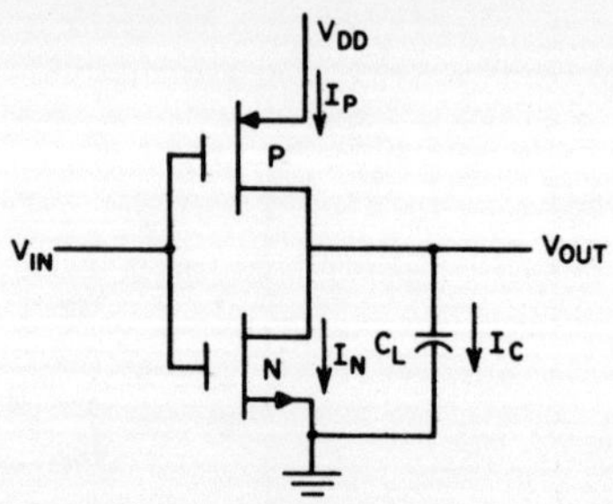

Figure 9.6. COS MOS inverter driving a load capacitance.

The switching response of the complementary inverter can be computed for an idealized case. Figure 9.6 illustrates an inverter driving a load capacitance, C_L. If the assumption is made that C_L is high in comparison to the device capacitance, the following approximate relationship can be derived for a stage delay, T_D:

$$T_D \cong 0.9\,\tau\left[1\Big/\left(1-\frac{V_{tn}{}^2}{V_{DD}}\right)+1\Big/\frac{K_p}{K_n}\left(1-\frac{|V_{tp}|}{V_{DD}}\right)^2\right] \tag{9.7}$$

where $\tau = C_L/(K_n V_{DD})$.

It can also be shown that for minimum delay and equal rise and fall times, the transconductance of the devices should be matched:

$$g_{mp} = g_{mn} \text{ if } W_p \simeq 2\ W_n.$$

This is the same transconductance value desired for maximum noise immunity.

Again referring to Figure 9.6, the actual switching response is essentially determined by the capacitance and the amount of current available from the transistors to charge this capacitance:

$$I_P = I_N + I_C$$

or

$$C\frac{dV}{dt} = I_P - I_N$$

and

$$T = C\int_{V_0}^{v}\frac{dV}{(I_P - I_N)}.$$

It is evident that minimum switching time is obtained when the current available to charge C for any voltage V is at a maximum.

9.3 CAPACITANCE

Source and drain capacitances depend on substrate bias and diffusion dopant as shown in Table 9.1.

TABLE 9.1 Source and Drain Capacitance per Unit Area

	Diffusion	
Bias (V)	$p+$ (pF/mil^2)	$n+$ (pF/mil^2)
0	0.12	0.2
10	0.03	0.05

Internal COS MOS capacitances are generally very small (in the neighborhood of 1 pF) and therefore devices on the chip that drive other devices can be very small.

9.4 COMPLEMENTARY NAND OR NOR GATES

Complex gates, such as three-input NOR or NAND gates, involve the use of series and parallel devices such as shown in Figures 9.7 and 9.8. When MOS devices are used in this manner, their individual characteristics must be compiled as conductances. If, in the inverter, the impedance of the 2-mil p-channel device is considered to be 1, then to provide the same

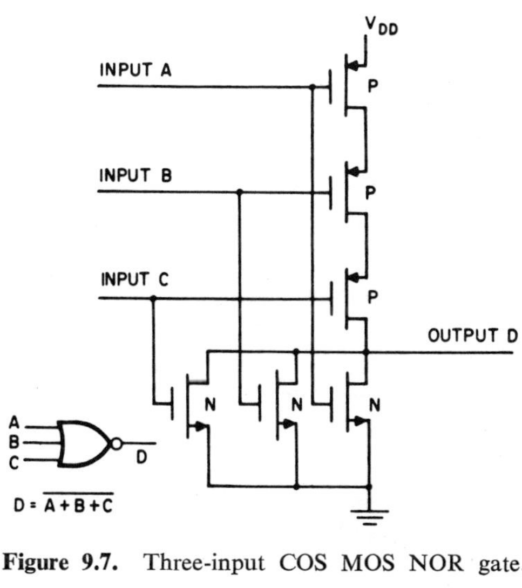

Figure 9.7. Three-input COS MOS NOR gate.

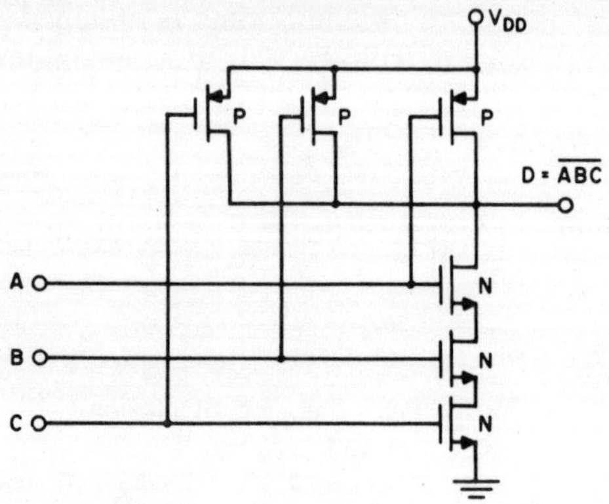

Figure 9.8. Three-input NAND gate.

impedance from the output to V_{pp} in the NOR gate, the three *p*-channel devices in series must also have a total impedance of 1; therefore each device will have an impedance of ⅓. The width of each will then be $3/1 \times 2 = 6$ mils. Since any one of the paralel *n*-channel units, if turned on, will bring the output to ground, each one must have the same impedance as the inverter *n*-channel devices; therefore, each should be 1 mil wide.

The switching response of even relatively simple circuits such as a two-input NAND gate cannot be readily calculated, since with series devices, the substrate-to-source capacitance and the threshold voltage are both functions of substrate bias, which is continually changing as the load capacitance is discharged.

9.5 SHIFT-REGISTER STAGE

A shift-register stage illustrating the reduction of components obtained by use of the capacitance storage feature of MOS gates is shown in Figures 9.9 and 9.10. In Figure 9.9, S_1 and S_2 are in series with the cross-coupling connections of the flip-flop. When these switches are opened, two separate storage elements are formed, the left side and right side of the flip-flop. Each side can store information for an extended but finite period of time. Digital information from the preceding stage is transferred to the left side through the gate and can change the stage on the left side. At the same time, the right side transfers its information to the left side of the next stage. If switch S_2 is closed first, and then S_1, the "1" or "0" information transferred from the preceding bit is locked into the left side. A shift register based on these principles and using COS MOS units for switching is shown in Figure 9.10; input data appear in Table 9.2. Essentially, this circuit is a simple two-phase MOS flip-flop.

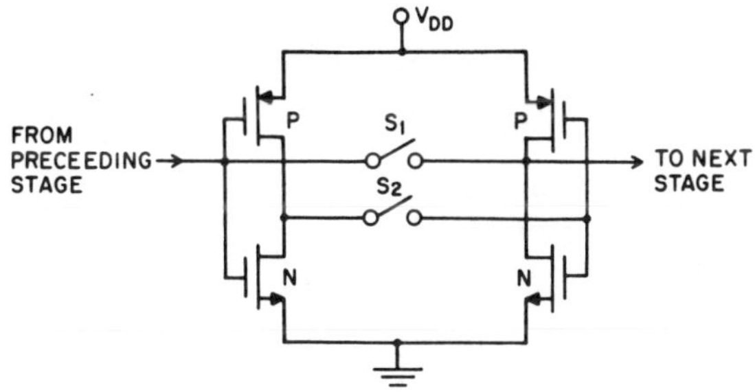

Figure 9.9. Shift-register stage.

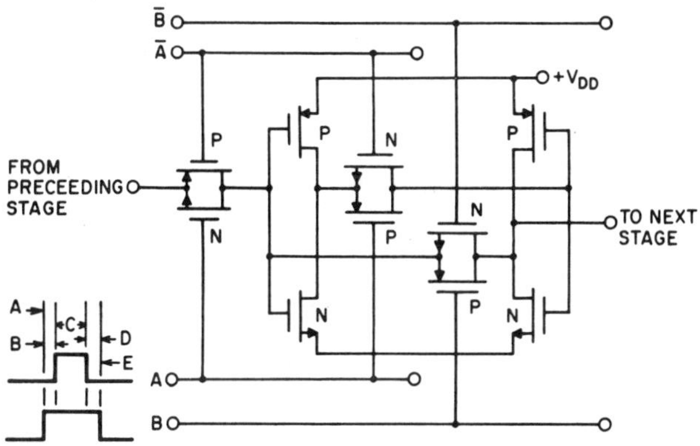

Figure 9.10. COS MOS shift-register stage.

Logic designs may be achieved by use of only NOR gates and inverters. However, a somewhat more economical gate-design approach is to consider (1) that common source and drain regions can exist in integrated-circuit form and (2) that the MOS transistor is topologically symmetrical.

TABLE 9.2 Shift-Register Input Data

Data Input Path	Flip-Flop Load	Operation
A Open	Closed	Information locked in flip-flop
B Open	Feedback open	First inverter freed to accept new data
C Closed	Both paths open	Data from preceding stage shifted into first inverter; data from output shifted to next stage
D Open	Feedback open	Data shifted dynamically from first to second inverter
E Open	Closed	Information locked in flip-flop

9.6 MEMORY CELL AND READ/WRITE DRIVERS

Figures 9.11 and 9.12 illustrate a ten-device basic memory cell and an associated read/write driver, respectively. The write driver is activated only upon a command common for the entire memory. An additional signal, the inverse of the write command, is also generated to ensure high-speed operation of the storage cell. These lines are then connected to appropriate points of the basic cell, shown in Figure 9.12. The cell can be written into by means of two MOS transmission gates—one for opening the feedback loop from *B* to *C*, the other for entering the desired information into the cell at *C* from the digit line. After one pair-delay time, the potentials at *B* and *C* are identical and the feedback loop may be closed by bringing the write line down to ground potential, thereby leaving the information stored in the flip-flop. Readout from the memory is accomplished nondestructively by controlling the conductance of an *n*-type MOS device with the output voltage of the cell. Subsequent application of a

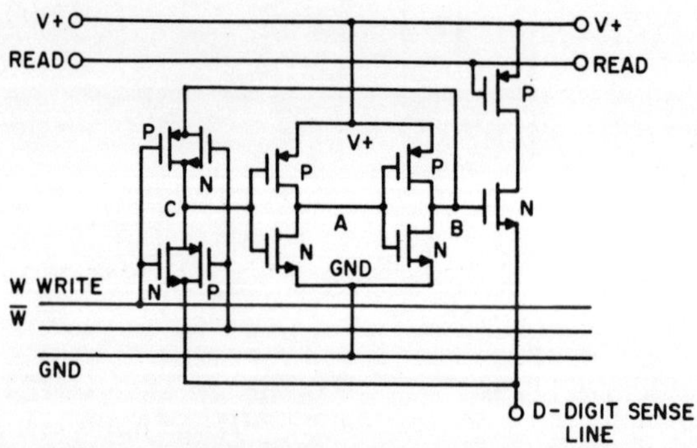

Figure 9.11. MOS memory cell with nondestructive readout.

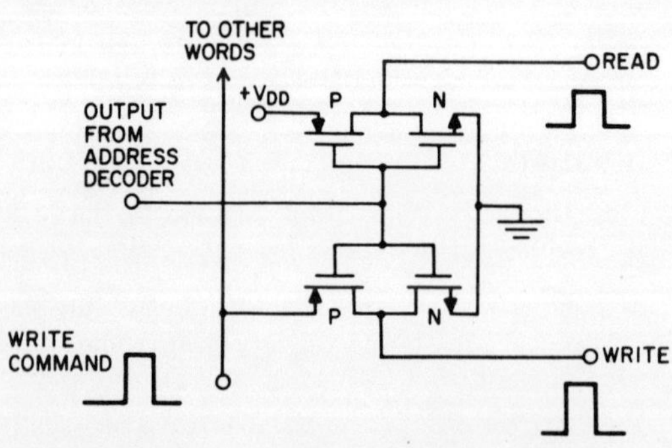

Figure 9.12. MOS read/write drivers.

read command will result in the presence or absence of a current in the low-impedance sense (digit) line, indicating a stored "1" or "0." Current sensing with a low-impedance device, such as a grounded-base bipolar transistor, is a very fast operation even if the capacitance of the digit line is large, as it would be in a large memory.

9.7 PAIR DELAY

The switching response of a COS MOS logic circuit is determined solely by capacitance loading at the input and output and the amount of current available to charge this capacitance.

One of the most important parameters for a logic inverter is the circuit-pair delay, which is defined as the time delay experienced by the signal in traversing two inversions as measured at the halfway points on the waveform.

Referring again to (9.7) and Figure 9.6, we can see that pair delay is a function of effective output loading capacitance, assuming that the input waveform is a square pulse.

The input capacitance of a COS MOS transistor is approximately:

$$C_{IN} = C_{GC} + C_{CS} + 2C_{GD} \tag{9.8}$$

$$\cong 0.135 \text{ pF/mil of channel width.}$$

The total effective C_L of an inverter stage driving n like stages is approximately:

$$C_L = n\,[C_{\mathrm{IN}\,(n)} + C_{\mathrm{IN}\,(p)}] + C_{\mathrm{OUT}} + C_{\mathrm{stray}}. \tag{9.9}$$

where $C_{\mathrm{IN}\,(n)}$ and $C_{\mathrm{IN}\,(p)}$ represent C_{IN} values for the n-channel and p-channel devices, respectively.

As a typical example, let us assume an inverter stage is driving five others ($n = 5$). The MOS device characteristics are given in Table 9.3. The supply voltage V_{DD} is 10 V, and

TABLE 9.3 MOS Device Characteristics

$W_N = 1$ mil	$C_{\mathrm{OUT}} \simeq 0.46$ pF
$V_{tn} = 2$ V	$t_{\mathrm{OX}} = 1000$ Å
$W_p = 2$ mils	$E_{\mathrm{OX}} = 35 \times 10^{-12}$ farad/meter
$V_{tp} = 2$ V	$K_N = 6 \times 10^{-6}\, W_N/L_N = 24 \times 10^{-6}$
$L_N = L_p = 0.25$ mil	$K_p = 3 \times 10^{-6}\, W_p/L_p = 24 \times 10^{-6}$
$C_{\mathrm{stray\ wiring}} = 1$ pF	
$C_L = 5\,[0.135 \times 1 + 0.135 \times 2] + 0.46 + 1 = 3.48$ pF.	

$$\tau = \frac{C_L}{K_N V_{DD}} = \frac{3.48 \times 10^{-12}}{24 \times 10^{-6} \times 10} = 14.5 \text{ ns}$$

$$T_D = .9\,\tau \left[1 \Big/ \left(1 - \frac{V_{tn}}{V_{DD}}\right)^2 + 1 \Big/ \frac{K_p}{K_N} \left(1 - \frac{V_{tp}}{V_{DD}}\right)^2 \right] \tag{9.9}$$

$$T_D = .9 \times 14.5 \times 10^{-9} \left[1 \Big/ \left(1 - \frac{2}{10}\right)^2 + 1 \Big/ \left(1 - \frac{1}{10}\right)^2 \right] = 41 \text{ ns.}$$

The above results are typical for COS MOS gates. The results, of course, are highly dependent on fanout and geometry. For example, a gate used to drive many others can be larger than those driven, thus reducing T_D.

9.8 POWER DISSIPATION

Power dissipation is an extremely important consideration in large-scale integrated circuits, and when large numbers of devices are involved, it may be the ultimate limitation in determining device packing density. Fortunately, the standby power dissipation of COS MOS circuits is only in the order of nanowatts. However, when a COS MOS circuit switches, a considerable amount of current flows during the switching interval. The average power dissipation during switching can be approximated by the following expression:

$$P = C_L V_{DD}^2 f \tag{9.10}$$

where f is the switching frequency. The total dissipation is given by:

$$P_{\text{total}} = P_{\text{static}} + C_L V_{DD}^2 f. \tag{9.11}$$

For example, if a 50-gate array is considered, the standby power will be approximately 50 nW (at $V_{DD} = 10$ V). If the switching rate is 1 MHz, and a total load capacitance of 50 pF is being driven, then:

$$P_t = 50 \times 10^{-9} + (50 \times 10^{-12})\,(10^2 \times 10^6)$$

or

$$P_t = 50 \text{ nW} + 5 \text{ mW} = 5.00005 \text{ mW}.$$

If bipolar devices had been used to provide the 50 gates, the second term of this equation would not have changed appreciably, however, the static term would be 8 to 10 mW per gate, and the total power dissipation would be

$$P_t = 50 \times 10\text{ mW} + 5\text{ mW} = 505\text{ mW}.$$

A *p*-channel MOS array using multiphase clocking might have a static power dissipation of 1 mW per gate, or $P_t = 55$ mW.

9.9 SPEED–POWER COMPARISON

Various combinations of speed, dissipation, and noise immunity are available with present bipolar, MOS, and COS MOS integrated circuits.

The speed–power–noise immunity relations among bipolar, MOS, and COS MOS integrated gates are shown in Figure 9.13. Curves 1 and 2 are for single gates (usually used to drive many other gates), and curves 3–5 apply for gates that are internal to arrays and drive other gates on the same chip. These curves reveal that COS MOS devices offer a considerable range of operating characteristics. They can be designed with gate thresholds from 1.5 V to approximately 7 to 8 V in various channel widths and lengths, and with power supplies ranging from 6 to 20 V. In addition, they offer extremely low-static power dissipation.

9.10 COS MOS ADVANTAGES

The area required to perform many logic functions in an integrated COS MOS array is about one-tenth that required with bipolar technology. COS MOS processing requires fewer steps (30 to 40% less) than are used in the manufacture of bipolar ICs and COS MOS costs are dropping rapidly, although selling prices are still higher than prices for bipolar devices. However, since COS MOS ICs are operated from a single power

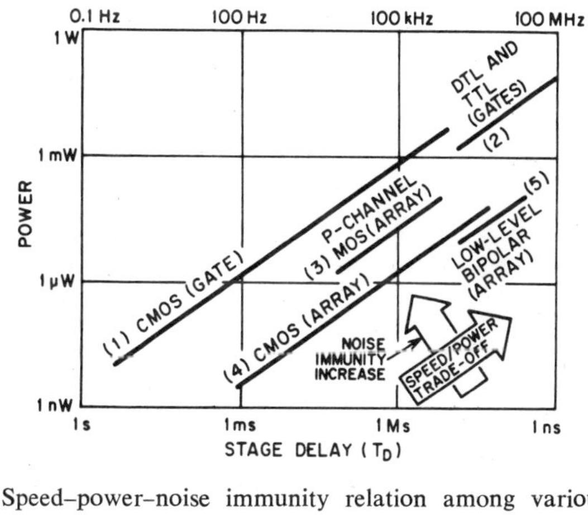

Figure 9.13. Speed–power–noise immunity relation among various types of integrated gates.

source, considerable flexibility is offered to system designers for speed–power optimization.

Obviously COS MOS cannot match the gate speed of ECL, TTL, or even DTL bipolar types; but size, system, cost, and cooling requirements can make COS MOS more favorable in moderate-speed applications.

9.11 SEQUENTIAL MEMORIES

Let's survey a number of applications in which the characteristics of COS MOS can be used to advantage. The shift register is one of the most important and useful elements in the arithmetic unit of computers. It is commonly used as a delay line for digital signals as well as for temporary storage registers.

A shift register has both a primary and a secondary storage element at each stage. The primary storage element holds the information. At a clocking command, this information is shifted to the secondary storage element of the next stage. At the next command, the information is shifted to the appropriate primary storage element. The two-step process eliminates timing problems by preventing coupling of the primary storage elements of successive stages. In a COS MOS IC, the primary and secondary storage functions may be combined either in a flip-flop or in a capacitor. An MOS gate provides the capacitance for temporary storage; a flip-flop permanent storage. If both primary and secondary storage are capacitive, an ac or dynamic shift register results.

9.12 SHIFT REGISTERS

An elementary dynamic shift-register stage is shown in Figure 9.14. During clock time T_1, gate N_3 is closed and gate P_4 is open; the reverse condition is obtained during T_2. The capacitance represented by P_1 and N_1, the gates of the first inverter, is the primary storage medium, and the capacitance of P_2 and N_2 is the secondary storage medium.

During T_2, the output inverter P_2 and N_2 is decoupled from P_1 and N_1 because N_3 is open. Gate P_4 is conducting, and information from the preceding stage can charge or discharge the capacitance of gates P_1 and N_1. Simultaneously, the data output provides an information transfer to the next stage.

When the clock becomes positive (T_1), P_4 opens and N_3 conducts, and information is shifted to the data output. Therefore, the cycle is completed within one clocking pulse.

Static shift registers offer permanent storage capability. A static shift-

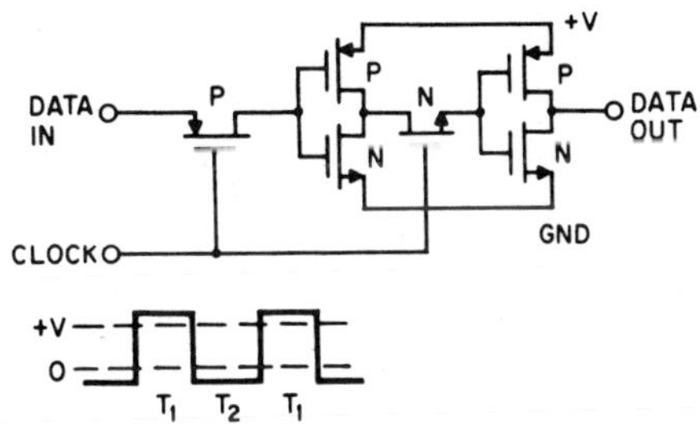

Figure 9.14. Elementary dynamic shift-register stage.

register stage using dual transmission gates was described in connection with Figures 9.9 and 9.10.

9.13 MEMORIES

The rapid development of IC technology has made the semiconductor memory an economical alternative for magnetic-core memories in capacities up to 4096 bits. The indications are that future high-speed scratch-pad and possibly main memories will be built entirely with large, monolithic semiconductor arrays.

The random-access storage matrix and the address decoders used to select a particular memory cell location can be processed simultaneously on a common substrate. Most active memory cells currently in use employ flip-flops to store binary information; logic gating is incorporated in the cell to ensure that information is written into, or read out of, only the selected memory location. To decrease cycle time, the state of the cell is usually sensed in a nondestructive readout (NDRO) mode.

Several COS MOS memory circuits and structures have been conceived, for example, the ten-device memory cell described in Figures 9.11 and 9.12. Word-organized random-access memory modules complete with address decoders and word drivers have been built.

As an alternative to word-organized structures, designs using coincident voltage of *XY* memory cells in a so-called bit-organized system are possible. In these memory cells the coincidence of a "1" in the *X* and *Y* lines is performed in the cell, and the rest of the decoding is done externally. In a system with N words, a word-organized memory would require N word drivers, whereas a bit-organized memory would require $2N/2$.

9.14 MEMORY COMPARISONS

Various forms of semiconductor and magnetic memories are compared in Table 9.4. A 10^6-bit memory is assumed for reference.

TABLE 9.4 Comparison of Semiconductor and Magnetic Characteristics for a Megabit Memory

	Readout	COS MOS (NDRO)	Bipolar (NDRO)	Core (DRO)	Plated Wire (NDRO/DRO)	Thin-Film (DRO/NDRO)
Unit Cell	Read delay (ns)	10	5	150	20	10
	Write delay (ns)	25	5	150	20	10
	Standby power (μW)	10	1000	0	0	0
Array of cells	Array size, bits	256 × 1	128 × 1	16 *k* × 18	4 *k* × 128	1024 × 64
	Word select 1 (mA)	1	20	425	900	200
	Digit write 1 (mA)	1	20	425	20	15
	Drive voltages (V)	5–15	2	30	20	10
	Readout signal	200 μA / Dc	500 μA / Dc	20–mV / Pulse	10–mV / Pulse	2–mV / Pulse
	Packing density (bpsi)	15 *k*	7.2 *k*	4.5 *k*	1450	3200
	Rise time of read current (ns)	—	—	100	30	10–35
	Curie temp. (C)	—	—	500–600	600	600
	Volatility	Yes	Yes	No	No	No
10^6 Bit Memory	Economic module, bits	2–32 *k*	1–8 *k*	256 *k*	256 *k*	64 *k*
	Power dissipation (W)	50	1000	100	200	200
	Full cycle time (ns)	125	50–100	500	200	200
	Cost per bit (cents) in mass production	2–5	4–5	1–2	2–4	2–3
	Mode of organization	LS*	LS	2–1/2D	LS	LS
	Batch fabrication	Yes	Yes	No	Semi	Yes

* LS = Linear select

10

Characteristics of Complementary MOS ICs

by Merle V. Hoover

RCA Corporation

The basic COS MOS inverter circuit is shown in Figure 10.1*a*. This monolithic circuit consists of one *p*-channel enhancement-type MOS transistor and one *n*-channel enhancement-type MOS transistor, with their drains tied in common and effectively operating in a series connection from a single power supply. The logic-swing gate voltage ranges from zero to the supply voltage, V. When the voltage of $+V$ (e.g., logical "1") is applied at the input, the *n*-channel unit is turned on and the *p*-channel unit is turned off; the output is then in the low state (logical "0"), and logic–voltage inversion has been accomplished, as illustrated by the curve in Figure 10.1*b*. When 0 voltage (e.g., logical "0") is applied at the input, the *p*-channel unit is turned on and the *n*-channel unit is turned off. This inverter action results in an output of $+V$ volts (logical "1"), as shown in the curve of Figure 10.1*c*.

It should be understood that in both logic states one unit is off while the other is on. Consequently the quiescent power dissipation is simply the product of the off-unit leakage current and the supply voltage. Because the off-unit leakage current is typically about 10 nA at a drain-to-source voltage of 10 V, the quiescent power dissipation for the simple inverter is typically about 10 nW in each logic state.

The input impedance of the typical COS MOS transistor pair is ex-

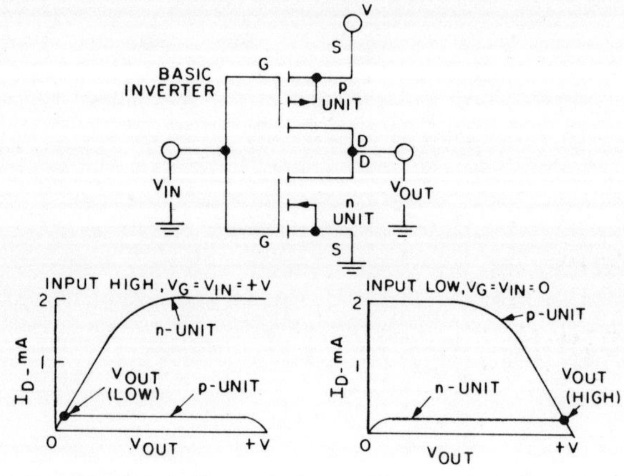

Figure 10.1. (*a*) Basic COS MOS inverter circuit; (*b*) and (*c*) typical operation currents.

tremely high, typically 10^{12} Ω. This high-input-impedance characteristic permits fanout of COS MOS pairs without significant increase of the low quiescent power consumption. Furthermore, there are essentially no dc restrictions on fanout capability (i.e., the circuits have infinite fanout capability). Although fanout does increase the capacitance loading, thus reducing speed, the practical fanout capability can be greater than 50.

During switching, when the capacitance in the output circuit (including the load) must be charged up to the "high" level through the "on" *p*-channel transistor, additional power is dissipated in the *p*-channel unit. Furthermore, the charge stored in the output and load capacitance is dissipated in the *n*-channel transistor when the input goes "high." The dynamic power *P* required for switching is given by

$$P = C_O V_{DD}^2 f$$

where C_O is the capacitance at the output node, V_{DD} is the supply voltage, and f is the frequency in hertz. Figure 10.2 shows the power dissipation as a function of frequency for the inverter circuit of Figure 10.1*a*. It is apparent that COS MOS circuits can be operated very efficiently because their quiescent power consumption is low and their dynamic power consumption is directly related to the switching rate required. In this respect, COS MOS circuits have a kinship with a magnetic core operating in a memory system because the core dissipates power only during each switching transient, but not in the quiescent "1" or "0" states.

COS MOS circuits have inherently sharp transfer characteristics, as

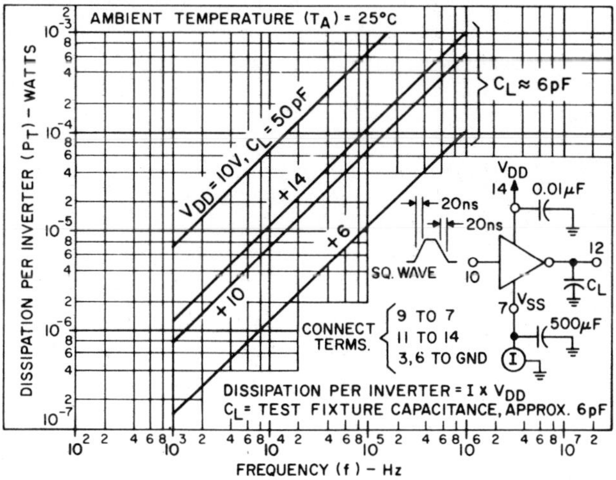

Figure 10.2. Power dissipation as a function of frequency for the inverter circuit of Figure 10.1*a*.

shown in Figure 10.3, for the simple inverter of Figure 10.1. As a consequence, the noise immunity of the logic circuits is typically about 40% of the power supply voltage (e.g., 4 V for a V_{DD} of 10 V). Figure 10.3 also shows the comparatively minor variations in transfer characteristics at the extremes of the operating-temperature range. COS MOS circuits make ex-

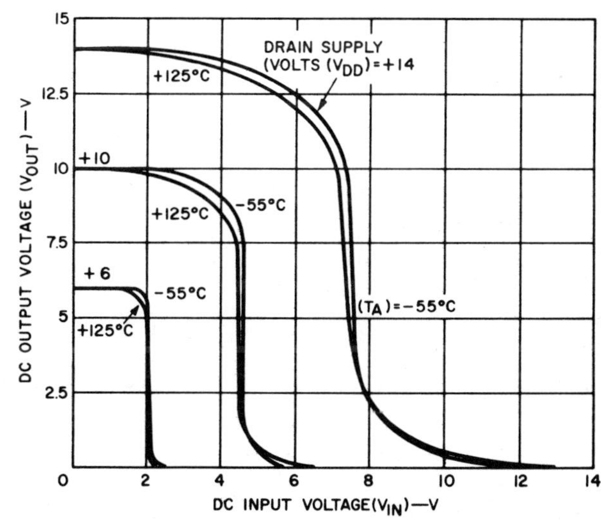

Figure 10.3. Variations in transfer characteristics for various temperatures.

cellent switches because extremely low saturation voltage (typically less than 1/mV) results in very large ratios of "off" to "on" output voltages. In addition they exhibit low "1" and "0" level output impedances, typically 1000 Ω with a 10-V supply.

10.1 COS MOS ARRAYS

Many circuits can be derived from simple COS MOS arrays. The array shown in Figure 10.4 is composed of three *n*-channel and three *p*-channel enhancement-type MOS transistors on a single, monolithic silicon chip. The COS MOS pair of transistors at the right side of Figure 10.4 is connected as an inverter in a configuration identical to that shown in Figure 10.1. A second inverter can be made available by interconnection of terminals 8 and 13, and a third inverter configuration can be provided by interconnection of terminals 1 and 5. This trio of inverters is one of numerous COS MOS logic-circuit arrangements that can be designed with the basic CD4007 circuit. Typical COS MOS logic-circuit arrangements are further illustrated in Figure 10.5.

10.2 DUAL *D*-TYPE FLIP-FLOP

Figure 10.6 shows the schematic diagram, logic block diagram, and truth tables for a dual *d*-type flip-flop. The circuit consists of two identical, in-

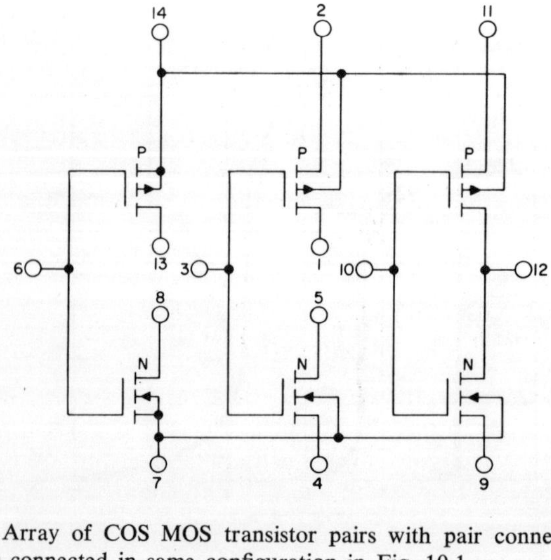

Figure 10.4. Array of COS MOS transistor pairs with pair connected as inverter; pair at right is connected in same configuration in Fig. 10.1.

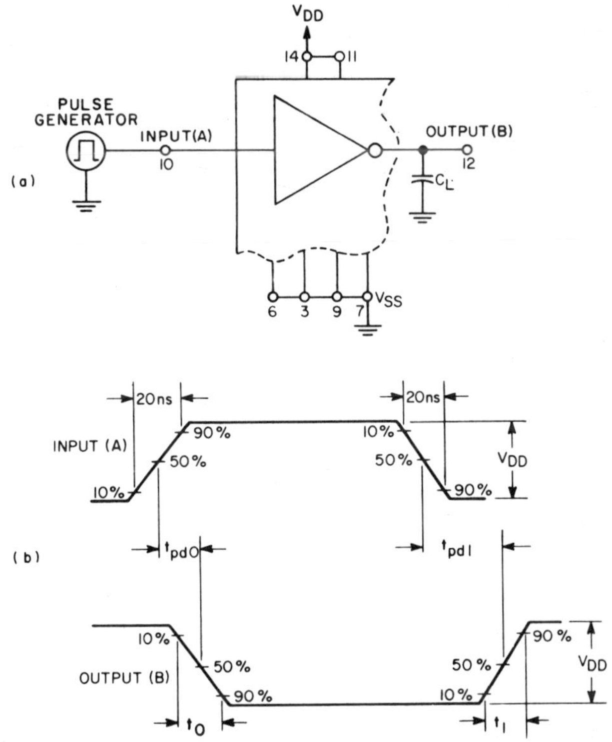

Figure 10.5. Typical COS MOS logic-circuit arrangements.

dependent, data-type flip-flops on an single, monolithic chip. Each flip-flop has independent data, reset, and clock inputs and complementary outputs. These devices can be used for shift-register applications and, by connection of the Q output to the data input, for counter and toggle applications. The logic level present at the D input is transferred to the Q output during the positive-going transition of the clock pulse. In static flip-flop operation, the circuit retains its state indefinitely, with the clock level either high or low.

10.3 COS MOS MEMORY

Diagrams for a 16-word, 1-bit, NDRO memory built with COS MOS transistors are shown in Figure 10.7. This circuit consists of 16 flip-flop storage cells arranged in an X–Y matrix for easy access, as shown in Figures 10.7*a* and 10.7*c*. Each storage cell contains a flip-flop, which consists of two cross-coupled COS MOS inverter stages and two transmission gates (also known as bidirectional gates). The schematic diagram of one COS MOS memory cell in Figure 10.7*c* shows the connection of two *n*-channel

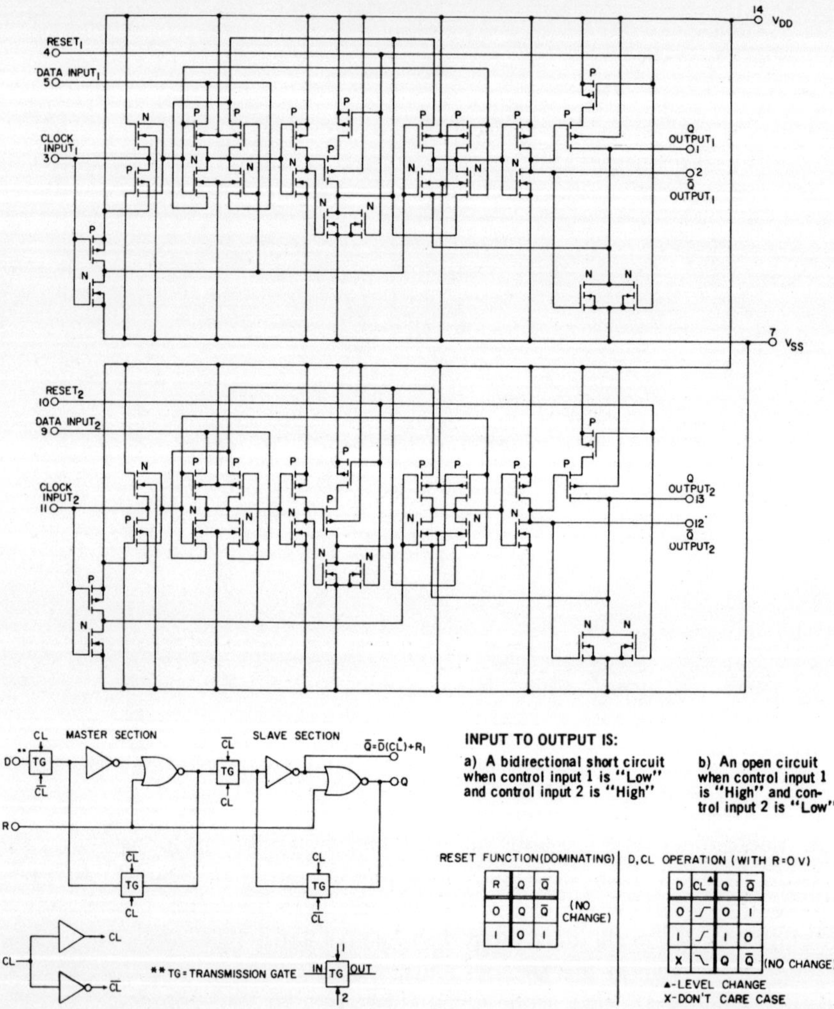

Figure 10.6. Circuit diagram, logic block diagram, and truth tables for dual D-type flip-flop COS MOS IC.

MOS transistors to perform the transmission gating to and from each cell. Figure 10.7*d* shows a truth table applicable to each cell.

10.4 MEMORY MODE OF OPERATION

Addressing of a particular cell is accomplished by the simultaneous application of high voltage levels on the X–Y coordinates of that cell. A "1"

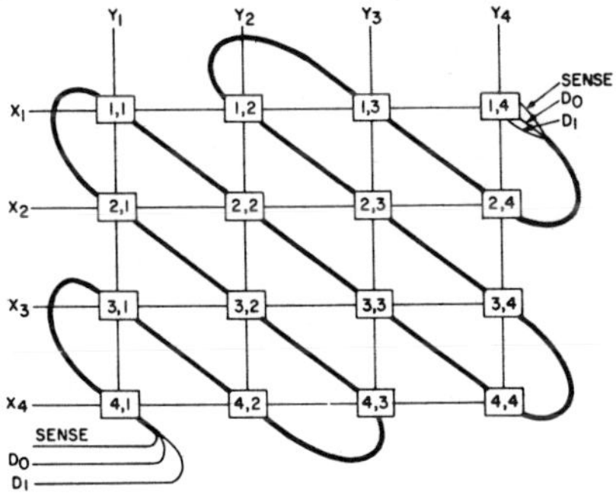

(a) Memory-Cel' Arrangement Diagram

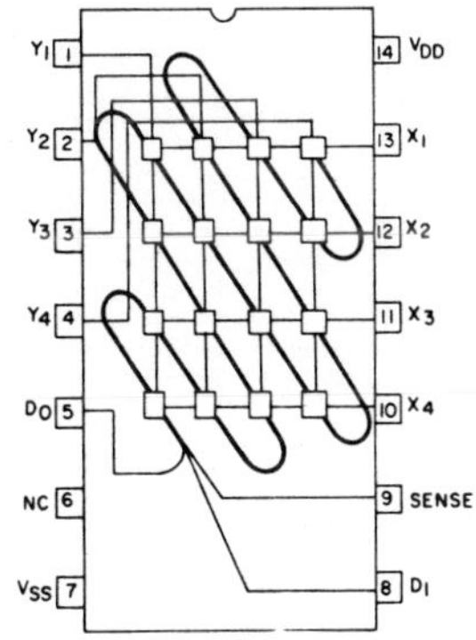

NOTE: ALL UNUSED INPUTS MUST BE CONNECTED TO THE MOST NEGATIVE CIRCUIT VOLTAGE (V_{SS}).

(b) Functional Diagram

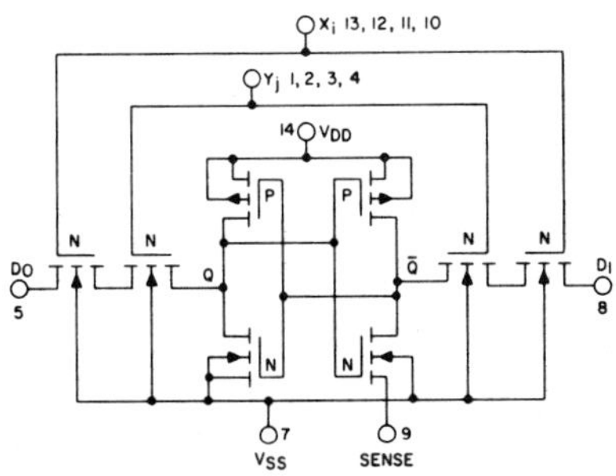

NOTE: V_{DD}, V_{SS}, D_0, D_1 AND SENSE LINES ARE COMMON TO ALL 16 CELLS.

X_i = X_1 OR X_2 OR X_3 OR X_4
Y_i = Y_1 OR Y_2 OR Y_3 OR Y_4

X AND Y LINES ARE COMMON TO FOUR CELLS EACH AND ARE CONNECTED IN MATRIX FORMAT.

(c) Schematic Diagram – One of 16 Memory Cells

WRITING VIA D_0 and D_1 LINES					
X	Y	D_0	D_1	CELL STATUS	SENSE CURRENT (mA)
0	0	X	X	NC	0
1	0	X	X	NC	0
0	1	X	X	NC	0
1	1	0	0	U	0
1	1	1	0	1	0
1	1	0	1	0	0
CURRENT SENSING AT "SENSE" LEAD					
1	1	1	1	1	0.5
1	1	1	1	0	0
LOGIC-LEVEL SENSING ON D_0 & D_1 LEADS					
1	1	Q	$\bar{Q}$	NC	"sense" line shorted to V_{SS}

POSITIVE LOGIC

Levels: "0" V_{SS}; "1" V_{DD}

X = DON'T CARE
NC = NO CHANGE
U = UNDETERMINABLE

(d) Truth Table (Each Cell)

Figure 10.7. COS MOS nondestructive readout memory: (*a*) memory cell arrangement diagram, (*b*) functional diagram, (*c*) schematic diagram of one of the 16 memory cells, (*d*) truth table (each cell), (*e*) typical timing diagram.

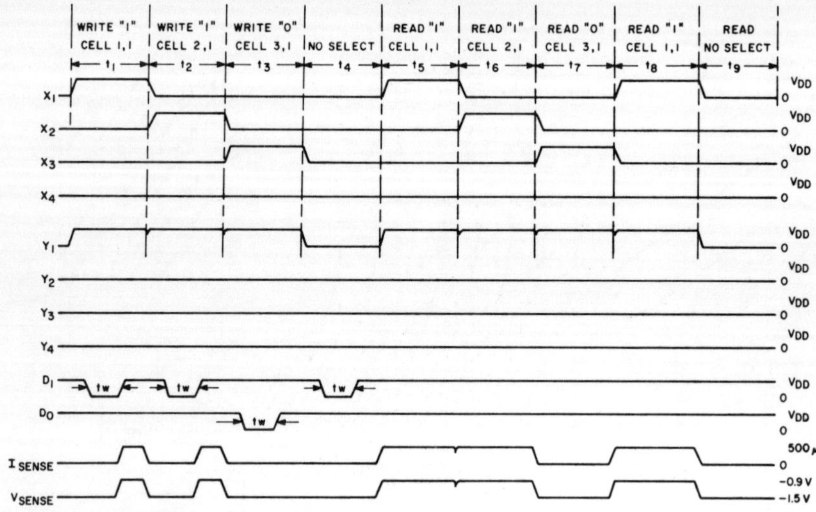

(e) Typical Timing Diagram

Fig. 10.7. (*Concluded*)

is written into an addressed cell by the application of a low voltage level on the D_1 line and a high voltage level on the D_0 line; conversely, a "0" is written into the addressed cell by forcing the D_0 line "low" and the D_1 line "high." Nondestructive readout or sensing at the $\bar{Q}$ sense-return line of an addressed cell is accomplished by providing sense current by means of an externally forced high-level voltage on the D_1 line. The D_0 line is also maintained at a high voltage level during readout.

Sensing can also be accomplished by external logic-level monitoring of the D_0 and D_1 lines. In this mode of sensing, logic circuitry must be provided that will inhibit the writing operation during readout, and the sense line must be connected to V_{SS}. This design permits direct logic-level addressing and writing and a choice of either logic-level or current-sensing readout. Addressing and sensing can be accomplished with MOS or bipolar transistor circuits. Figure 10.7*e* shows a typical timing diagram for the COS MOS memory circuit.

10.5 EXPANSION OF MEMORY

Expansion of the NDRO memory beyond the 16-word, 1-bit capacity can be accomplished easily by use of any number of CD4005 or CD4005D packages. Figure 10.8 shows how 32 packages can be interconnected to produce a 64-word, 8-bit memory. Word capacity is increased by horizontal expansion in the X and Y directions (increasing the number of cell lo-

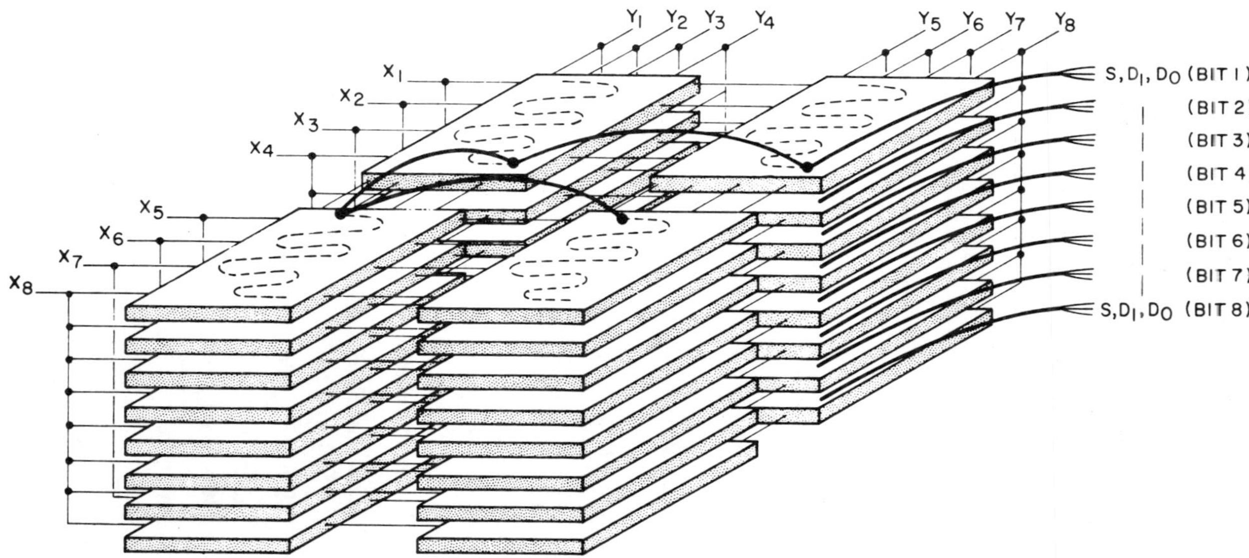

Figure 10.8. COS MOS 64-word, 8-bit memory assembled from 32 arrays of the type shown in Figure 10.7.

cations). Each plane represents one bit, and the sense lines in each plane are made common either by direct connection or through gates; similarly, the D_0 lines are made common, and the D_1 lines are made common in the same manner.

Bit expansion is accomplished by an increase in the number of horizontal planes. The corresponding X and Y terminals in each plane are connected together directly or through gating.

Consider, as an example, writing the binary word 10111011 (least significant bit is bit 1) into word locations 3, 4:

1. X_3 and Y_4 inputs are set to a high level (H).

2.

Bit Number	D_0	D_1
1	H	L
2	H	L
3	L	H
4	H	L
5	H	L
6	H	L
7	L	H
8	L	H

To read the same word out of word location 3, 4:

1. X_3 and Y_4 are again set to a high level (H).
2. Current from an external source is supplied on the D_1 line for each bit (D_1 is set to a high level).
3. All D_0 lines are maintained at a high voltage level.

All sense leads associated with cells in which a "1" was stored will carry approximately 500 μA; cells containing a "0" will pass virtually 0 current through their respective D_1 leads.

10.6 COS MOS SHIFT REGISTER

Figure 10.9 shows several diagrams for an 18-stage static shift register using COS MOS transistors. A functional analysis of the IC will reveal that the RCA CD4006 or CD4006D is composed of four separate "shift-register" sections; two sections contain four stages and each of the other two sections has five stages (4 + 1). Each section has an independent data input and a Q output from the fourth stage. This configuration permits easy implementation of various shift-register lengths in the range from 4 to 18 stages. Specifically, multiple-register sections of 4, 5, 8, and 9 stages or

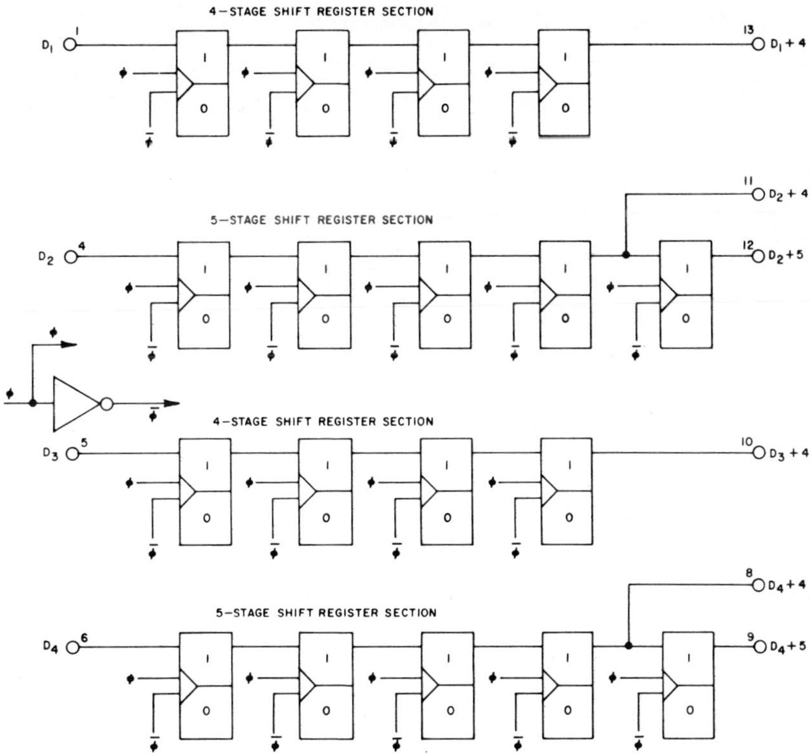

Logic Diagram

D	ϕ▲	D + 1
0	⌉_	0
1	⌉_	1
X	_⌈	NC

NC = No Change;
X = Don't Care;
▲ = Level Change

Truth Table for One Shift Register Stage

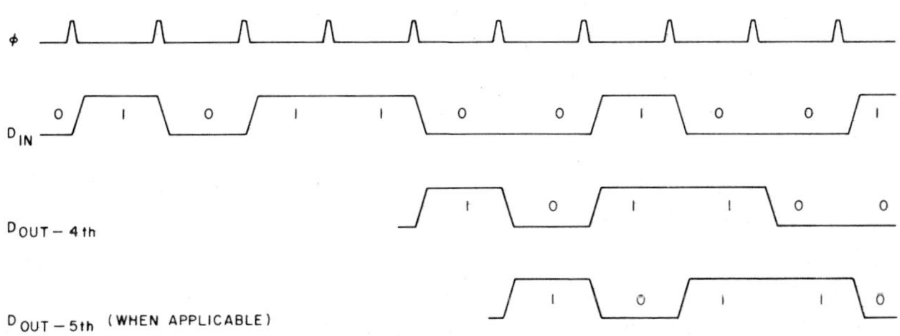

Typical Tuning Diagram for One Register Section

Figure 10.9. Logic diagrams, truth table, and timing diagram for an 18-stage shift register using COS MOS transistors.

single-register sections of 10, 12, 13, 14, 16, 17, and 18 can be implemented by use of one CD4006 or CD4006D package. Longer shift-register sections can be assembled by use of more than one package.

As might be expected, a single COS MOS shift-register stage is in reality a flip-flop coupled at the input and the output by transmission gates, [1–4].

A common clock signal is used for all stages. Information presented at the *D* line of any of the four register sections is shifted into that particular section under the control of the common clock signal. A "1" or a "0" logic level at the *D* line during the negative-going transition of the clock signal is transferred into stage 1 of the shift-register section, as shown in Figure 10.9. Simultaneously, the information in all the register stages shifts over one position. The fourth and fifth stages of a register subsequently present the *D* input information, delayed by four and five clock periods, respectively. Circuit operation is such that permanent register storage exists for a low clock-line state. The clock-line high level, however, should be less than 10 μs wide to assure proper register operation.

10.7 TRUE COMPLEMENTARY-LOGIC SHIFT REGISTER

COS MOS transistors can be used to form a true complementary-logic shift register consisting of two flip-flops per bit for two-phase operation [5]. However, a significant reduction in components can be obtained by the use of the excellent capacitance-storage features of COS MOS gates. As shown in Figure 10.10, switches S_1 and S_2 are in series with the cross-coupling connections of the flip-flop. When these switches are open, two separate storage elements are formed, i.e., the left side and the right side of the flip-flop. Each side can store information for an extended (but finite) period of time. Digital information from a "preceding stage" is applied to the gates of the left side, and can thereby change the state of the left side. Simultaneously, the right side can be imparting its information to the left side of the next stage. If switch S_2 is closed first, followed by S_1, the "1" or "0" information transferred from the preceding bit is locked into the left side.

If the switches S_1 and S_2 in Figure 10.10 are replaced by MOS devices and appropriate clocking connections, the static shift-register circuit shown in Figure 10.11 can be derived. Transistor N_5 locks the flip-flop comprising P_1–N_1 and P_2–N_2. However, because information must flow from left to right when N_3 and N_5 conduct, N_5 is designed to conduct much less current than N_3 when equal gate voltages are applied. As a result, the capacitance represented by the gates of P_1–N_1 charges much more slowly than the capacitance of gates P_2–N_2, and information is propagated from left to right. Clock voltage levels can be made equal to logic levels in this circuit.

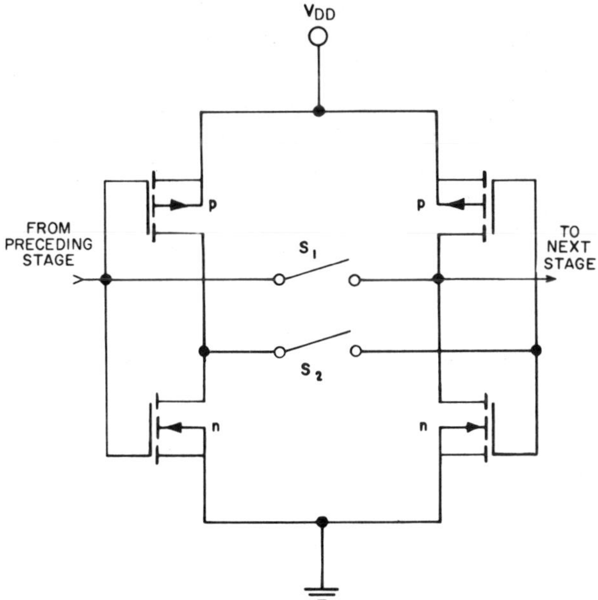

Figure 10.10. Complementary-logic shift-register stage.

The dual-transmission-gate version [6] of the circuit in Figure 10.10 is shown in Figure 10.11. COS MOS transistors P_1–N_1 and P_2–N_2 comprise the flip-flop; the COS MOS transmission gates P_3–N_3 and P_5–N_5 are analogous to the switches S_1 and S_2 of Figure 10.10. Another transmission gate, P_4–N_4, provides clock-controlled data input. The COS MOS inverter,

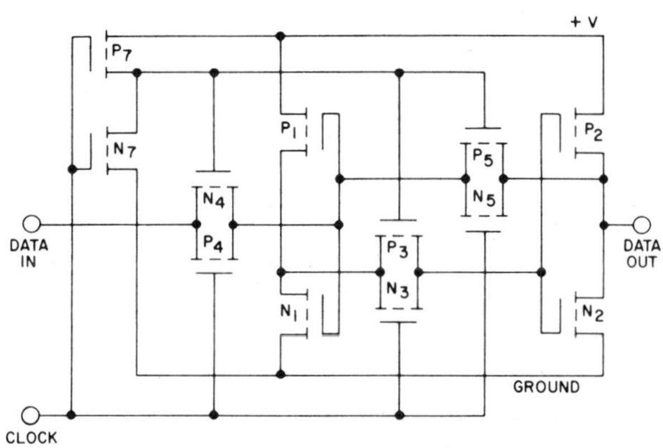

Figure 10.11. Dual-transmission-gate version of the circuit shown in Figure 10.10.

P_7–N_7, provides a phase reversal of the clock signal to actuate the appropriate sections of the transmission gates in the circuit.

The circuit of Figure 10.12 uses two gate-controlled flip-flops. Information is gated into the fiip-flops through a transmission gate, P_5–N_5 or P_8–N_8, and is locked by the action of P_6–N_6 or P_7–N_7. This circuit can be used as a binary counter if the inverted data output is converted to a data input; the clock line then becomes the binary input. Forms of these circuits are used in the COS MOS flip-flop circuit (Figure 10.6).

10.8 COS MOS LSI

Large-scael-integration versions of COS MOS should be along soon because of the desirability of the following features:

1. The relatively small chip area consumed by MOS transistors and the elimination of area-consuming resistors result in high circuit density per unit of silicon area.
2. The low quiescent power consumption makes it feasible to design comparatively complex arrays on a single chip without encountering restrictions caused by power dissipation.
3. COS MOS circuits are suitable for operation over the military temperature range of -55 to $+125$°C; furthermore, the inherent symmetry of the COS MOS configuration minimizes parameter variations as a function of temperature because there is a degree of inherent temperature compensation.
4. COS MOS technology permits the use of simple "tunnel interconnections" in complex arrays; the design of equivalent complexity in a bipolar LSI array frequently requires the use of "multilayer metalization" interconnections with additional processing steps and their attendant costs.

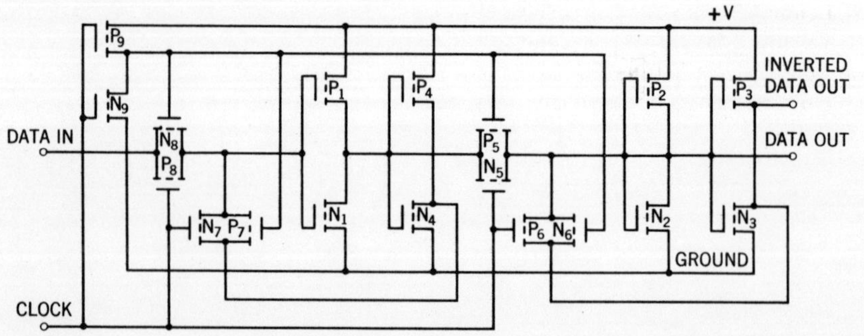

Figure 10.12. Static shift-register stage with dual transmission gates and single-phase clock.

5. COS MOS circuits have high noise immunity.

6. The circuits can operate from single power supplies of positive or negative polarity over a wide range of unregulated supply voltages.

7. Operation is reliable despite wide variations in transistor characteristics.

8. COS MOS transistors can easily be paralleled for size variation.

9. Simple circuits can be designed for single-phase clock operation in the range from drain current to frequencies of several megahertz.

10. Fixed logic-level swings are relatively independent of system tolerances and fanout.

11. Significant reductions in power consumption can provide solutions to heat-dissipation problems in compact design and/or promote the conversion of line-connected equipment to portable, battery-operated versions.

12. The available technology is capable of economically reproducing stable COS MOS units.

13. COS MOS transistors do not require the use of biasing networks.

14. Logic-function realization is simplified.

REFERENCES

[1] Wood, J., and Ball, R. G., "The Use of Insulated-Gate Field-Effect Transistors in Digital Storage Systems," *International Solid-State Circuit Conference Digests,* February, 1965.

[2] Burns, J. R., Gibson, J. J., Hu, K. C., and Powlus, R. A., "Integrated Memory Using Complementary Field-Effect Transistors," *International Solid-State Circuit Conference Digests,* February, 1966.

[3] Borkan, H., Rapp, A. K., Wennik, L. P., and Keller, K. R., "Complementary MOS Integrated Binary Counter," *International Solid-State Circuit Conference Digests,* February, 1967.

[4] Medwin, A. H., and Zuk, B., "Using MOS Transistors in Integrated Circuits," *Electronic Design,* November 23 and December 7, 1964.

[5] Lohman, R. D., "Some Applications of Metal-Oxide Semiconductors to Switching Circuits," *Semiconductor Products and Solid State Technology,* May, 1964.

[6] Ahrons, R. W., "MOS Complementary Shift Registers and Counters," *Digests* of Government Microcircuits Applications Conference, October, 1968.

11

MOS Application Areas

by George Stevenson

General Instrument Microelectronics Limited

MOS integrated circuits can be used to advantage in varying degrees in almost all parts of a typical logic–memory system. A typical system might consist of four basic units: the interface unit, the arithmetic or function unit, the control unit, and the memory unit. Let's examine each of these units.

11.1 THE INTERFACE UNIT

The purpose of the interface unit is to ensure signal compatibility between the system and the outside world. In the past there has been difficulty in interfacing between MOS and bipolar integrated circuits. However, now MOS ICs are being designed for compatibility between DTL and TTL. MOS can now be driven from DTL or TTL, and vice versa. This allows users to get the best of both MOS and DTL–TTL worlds. Where MOS is a relatively poor performer—for example, in high-speed portions of a system or where power driving is required—it pays to use a bipolar integrated circuit. In the remainder of the system, MOS can be employed in appropriate spots so that each type of circuit can be exploited without having to use special interface circuits.

Let's take a look at the input characteristics of the interface unit. In general the input to an MOS IC looks like a reverse biased diode. Thus the current requirements are negligible and the device can be considered to be voltage operated.

One common misconception is that an MOS IC input, because of its high input impedance, is inherently noise sensitive. In practice, the IC's input is directly coupled to the output of the previous stage. Thus the impedance seen by the noise signal is the parallel combination of input and output impedance, and this is generally less than 10kΩ. Experience has shown that MOS integrated circuits can work efficiently in very noisy environments, such as alongside induction motors.

The output from an MOS integrated circuit has two important characteristics. The first is that it is a voltage output principally suited to driving other MOS circuits. The second involves the fact that an MOS IC can be operated as a voltage-controlled resistor.

These characteristics allow MOS ICs to drive other circuits requiring only a few milliamperes of current; if greater currents are required, then an output buffer can be provided.

11.2 THE ARITHMETIC FUNCTION UNIT

The purpose of this unit is to perform some operation on data. Here the two prime considerations are speed and cost. If speed is the more important factor, then a bipolar solution is probably called for. If cost is the main parameter, then MOS is undoubtedly the type of circuitry to use.

It is unusual with MOS to provide a specific add or subtract function. Generally this operation is so straightforward that it is provided wherever it is required. This is an important point, since with bipolar circuitry it is common to provide a general add/subtract unit which is time multiplexed to meet the various needs in the system. This method obviously complicates the control logic, and it also means that uncomfortably high circuit speeds must be used to achieve a lower overall system speed. With the MOS approach, arithmetic units are provided as required, thus simplifying the control logic and making circuit speed and system speed fall more closely into line. It is important not to confuse circuit speed and system speed.

MOS techniques are currently proving very useful in counting and dividing applications. There are currently circuits that perform bidirectional decade counting with binary-codes decimal (BCD) output and presetting facilities. Long divider chains giving many decades of division are also easily realized with MOS. One of the simplest MOS circuits is the shift register, and there are a great number of uses to which it can be put; for example, long divider chains or random number generations can be readily formed using feedback techniques with shift registers. Clock frequencies in this application can go up to 5 MHz. In addition to straightforward arithmetic functions, there is generally a need for more sophisticated functions. Again, shift-register techniques can be used to good effect. For example it is easy to construct a digital differential analyzer using these techniques, and the element can then be put to other uses, such as providing exponential and logarithmic functions.

A read-only memory can be programmed to provide any required function. This is usually done by considering the ROM to be equivalent to a look-up table. Thus square root functions, sine–cosine functions, and logarithmic functions can be readily performed. One interesting aspect of this device is that it provides a means of performing functions at high speed, for although circuit speed is slower than that obtainable with bipolar circuitry, there are no subroutine or carrying delays. Thus there is no need for a sophisticated speed-up function such as a look-ahead carrying system.

In conclusion, MOS can be used in an arithmetic or function unit. Its speed is slower than that provided by bipolar circuits, but this can often be offset by exploiting the increase in parallelism that is possible without cost penalty with MOS ICs.

11.3 THE CONTROL UNIT

The purpose of the control unit is, of course, to control and regulate the flow of data in the system. The control unit is generally the hardest unit to define logically, since its function is frequently both complicated and complex. This area, also, is most likely to contain errors or to be changed as the system is designed. Little can be said about the control unit because there is seldom an obvious pattern to its design. Where the design is fluid, it probably pays to use standard logic elements that can be easily changed. However, if the control logic unit is required in considerable quantity, it is well worth considering a special or custom MOS circuit to do the job.

One part of the control unit does occur with some frequency and this is the addressing or routing function; it pays to use MOS ICs in this area. The matrix structure of a decoder is particularly suited for MOS. MOS can be used to form multiplexors or commutators, which can handle either analog or digital data. One interesting point is that with MOS it is often feasible to use an address highway and perform the decoding on the circuit that is being addressed. This technique reduces the number of wires and interconnections in a system. The read-only memory is also of use in the control unit, since it is possible to program a particular sequence of events on the chip so that by using a counter to drive the ROM, the entire program can then be run through either in sequence or at random.

11.4 THE MEMORY UNIT

The purpose of the memory unit is to store data at one time for subsequent use at a later time, and here MOS shows the greatest promise.

In general storage units can be split into two categories: sequential access and random access.

In sequential-access systems such as the magnetic tape unit or delay line unit, data can only be accessed in some particular sequence. The MOS equivalent to the delay line is, of course, the shift register. Shift registers can be accessed in a cyclic fashion using the clock as a reference. The clock frequency currently in use goes from dc to above 5 MHz. This structure is particularly good as far as cost is concerned because a very large logic content is used with very few input and output connections.

In the random-access case the best-known memory unit is the core store. Here data can be read in and read out into any location in the store according to the address of that location. The MOS memory is analogous to the core store in that data can be both read in and read out. This is a significant achievement with a semiconductor device since, again, the excellent ratio of logic content to the number of leads keeps the cost per bit down. MOS storage systems are now competitive in price with magnetic-core stores. This is particularly true for small storage systems, because magnetic-core stores are generally attractive only when storage is in excess of 2000 bits of data. Another advantage of the semiconductor store is the compatibility it offers between general processing logic and the storage unit. This gives the systems designer much greater freedom.

12

Nitride MOS Integrated Circuits

by Colin Clifford

General Instrument Microelectronics Limited

The need for devices that can accept low-level voltage drive from TTL and DTL and also meet the high current requirements of TTL and DTL inputs has been obvious. The easiest way to meet this need is to reduce the threshold voltage of the MOS structure. However, although a number of compatible processes have been proposed, only two have a history of production; these are 1-0-0 orientation silicon and the silicon-nitride process.

21.1 THRESHOLD VOLTAGE

Threshold voltage is dependent on a number of process variables as shown by the equation:

$$V_{\mathrm{th}} = \phi_{\mathrm{ms}} - \frac{1}{C_o}\int_o^{Xo} \frac{X}{X_o} p(X)dX - \frac{Q_{SS}}{C_o} - \frac{QB}{C_o} + 20_{fN}. \tag{12.1}$$

The two significant terms in this equation for silicon-nitride process are Q_{SS}/C_o and QB/C_o. The charge per unit area in the depletion region is constant and equal to Q_{SS}, which is the surface state charge per unit area; C_o is the capacitance per unit gate area which, if altered, changes the threshold. In gate regions, for a given dielectric constant, the value of C_o

(and hence threshold voltage) is determined by the thickness of the dielectric. In other areas of the circuit it is essential that the highest voltages on the chip do not form transistors under the thick field oxide; this is called field inversion. Figure 12.1 shows the variation of threshold voltage with Q_{SS} and gate thickness (i.e., C_o) and Figure 12.2 reveals the variation of field inversion voltage with the same parameters.

If 1-1-1 orientation silicon with Q_{SS} of 2.5×10 in. Q/cm^2 is used, a gate thickness of 800 Å is required for a reasonable threshold voltage. This gate thickness however, is not a practical value for reliable operation and good yields. An equivalent electrical thickness of 800 Å of silicon dioxide can be achieved with a mechanically thicker dielectric.

Silicon nitride offers a dielectric constant of $E_N = 7.5$ as compared with an E_{QX} of 3.9 for silicon dioxide. Problems arise when silicon nitride is used by itself as the gate dielectric; transistors exhibit a hysteresis in their characteristics. This can be solved by using a composite construction of silicon dioxide and silicon nitride as shown in Figure 12.3. This results in an equivalent electrical thickness of 800 Å of silicon dioxide but a mechanical thickness of 1000 Å. An incidental but useful feature of the composite gate is that the occurrence of pin holes in the gates is considerably reduced, since the probability of a pinhole in the oxide lining up with one in the nitride is statistically remote. In this way a 2 V threshold is achieved with a

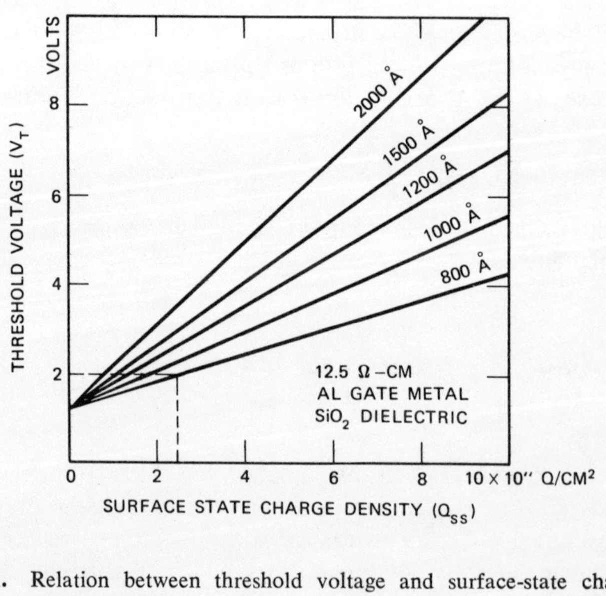

Figure 12.1. Relation between threshold voltage and surface-state charge density for various gate thicknesses.

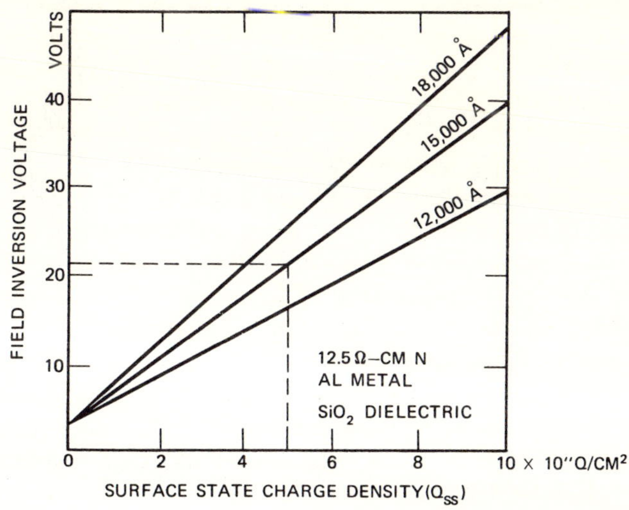

Figure 12.2. Relation between field inversion voltage and surface-state charge density for various gate thicknesses.

sufficiently high Q_{SS} to ensure a safe field inversion voltage. This is shown in Figure 12.2 to be about 22 V with a field oxide thickness of 1500 Å, which represents an adequate safety margin for circuits working between +5 and −12 V. This "sandwich" gate construction requires no more than the usual number of masks.

Another important parameter is the current-sinking capability at the output. In the circuit of Figure 12.4, transistor Q_2 sinks the I_{OL} current of the TTL gate; the transistor's current equation is given by:

$$I_D = K' \frac{W}{L} \left[(V_{GS} - V'_{th}) V_o - \frac{V_o^2}{2} \right]. \qquad (12.2)$$

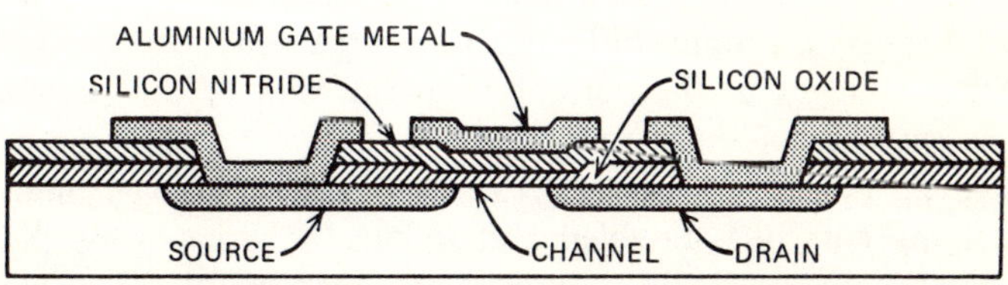

Figure 12.3. Field-effect-transistor construction using silicon-nitride and silicon-dioxide dielectric.

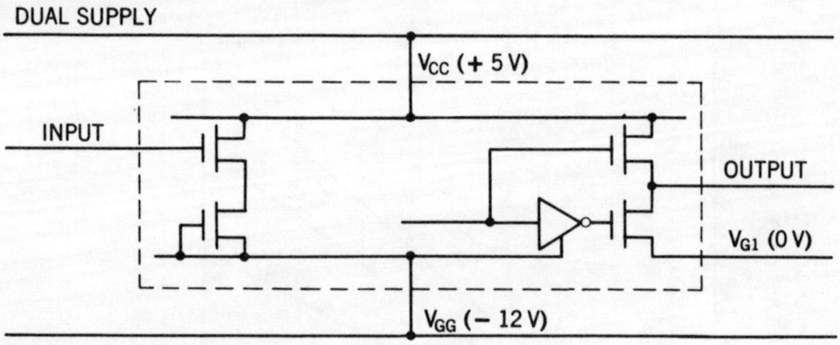

Figure 12.4. Basic circuit.

For a given set of voltages $I_D \propto K'\ (W/L)$ where K' is the gain factor and W/L is the geometric size of the transistor.

The gate dielectric is a composite structure, and the value for C_o is calculated taking into account the thickness and dielectric constants of the nitride and oxide as follows:

$$C_o = E_o \left(\frac{E_{ox}\, E_n}{t_n\, E_{ox} + t_{ox}\, E_n} \right). \tag{12.3}$$

Because of the higher permittivity of nitride E_N, the value of C_o, is increased and, since K' is equal to μC_o, the value of K' is higher than it would be with transistors using only a silicon-dioxide gate dielectric. This results in significant reductions in the size of the transistors required to sink current, with accompanying savings in chip area, particularly at the outputs.

12.2 NITRIDE ADVANTAGES

In addition to providing low threshold voltages, the nitride process provides a gain factor K' which is double what it would be with conventional processing. It has been shown that this increase in K' is obtained because of an increase in C_o, the capacitance per unit area of the transistor gate section; therefore the increase in gain is matched by the increase in C_o. Figure 12.5 shows a simple shift stage where the voltage at the gate of the second stage changes as the current from the load of the previous stage charges its gate capacitance.

There might appear to be no speed advantage for a given input voltage if the charging current is doubled and still has to drive twice the capacitance. Actually, the circuit diagram does not show the whole story; if reference is made to the composite of the layout of this stage on the chip (Figure 12.5), it can be seen that the gate capacitance is only part (about 8%) of the total capacitance, which is made up of the track and diffu-

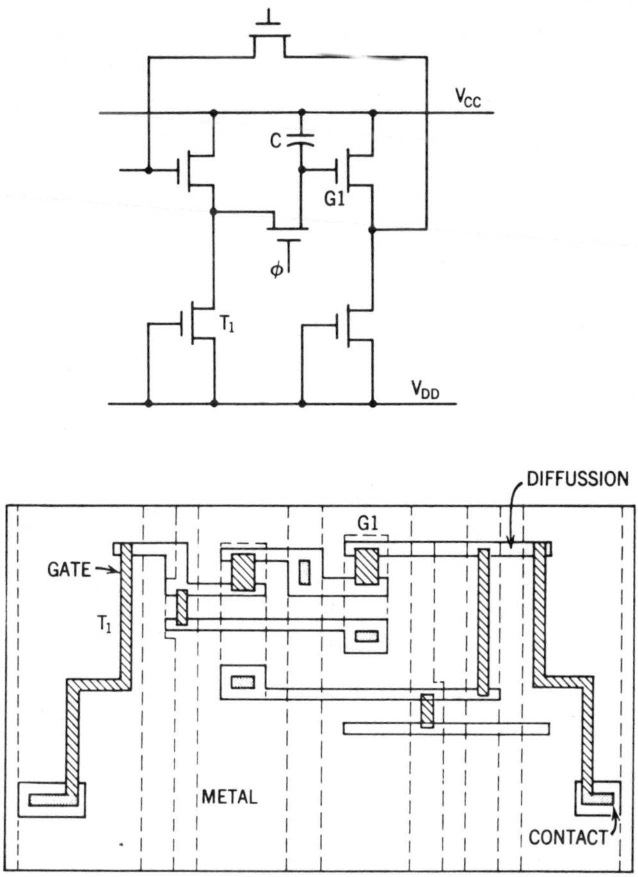

Figure 12.5. Simple shift-register stage. Layout diagram shows location of components.

sion capacitances. Thus if this capacitance is doubled, the total capacitance increases by only 15% even though K' is twice the value it exhibits in silicon-dioxide gate dielectric transistors. Thus a considerable increase in speed results.

Since the other major problem in producing TTL compatible circuits is sinking I_{OL} from the bipolar circuit, let us consider the output stage in Figure 12.4. This stage is chosen because of its low power dissipation and good drive capability and because it needs no external components. Clearly Q_1 has no difficulty in sinking I_{OH}; the key characteristic is the current-sinking capability of Q_2. The current equation for Q_2 is:

$$I_D = K' \frac{W}{L} \left[(V_{GS} - V'_{\text{th}}) V_o - \frac{V_o^2}{2} \right] \tag{12.4}$$

which for nominal supply voltages of +5 and −12 V at $V_{th} = 2$ V becomes:

$$I_D = K' \frac{W}{L} (3.5).$$

To find the geometric size of Q_2, $I_{OL} = 1.6$ mA is substituted:

$$\frac{W}{L} = \frac{4.6 \times 10^{-4}}{K'}.$$

It is clear that the value of K' will determine the W/L ratio; the higher the value of K', the smaller the size of the transistor. The advantage of a high value of K' is apparent; substituting a value derated for 125°C operation, transistor Q_2 has $W = 30{,}000$; if a conventional gate dielectric were used, Q_2 would obviously be twice as large. This leaves room for little else on the chip.

In an effort to reduce the size of the IC to compensate for processes with a low K' the drain of Q_2 can be connected to the negative supply rail V_{DD} instead of 0 V and the current equation will be:

$$I_D = \frac{K'}{2} \left(\frac{W}{L}\right) (V_{GS} - V_{th})^2. \tag{12.5}$$

Now a smaller transistor can be used. The transient response will require a larger transistor than the dc conditions indicate. This is because, in a low-frequency condition, the transistor can sink more current than the TTL demands and, as the output is clamped by the TTL substrate diodes, a current of typically 8 mA flows through Q_2. Because there is now 12 V across the transistor, the power dissipation in Q_2 is in the order of 100 mW. This has serious repercussions in that, for given package and operating temperature range, the power dissipated on the chip has a limiting value. For example, with a TO-5 can operated at 125°C, the chip power is limited to 300 mW, and if the circuit requires three TTL outputs there is no power available for logic.

In using nitride, advantage can be taken of the area and power savings of the low-power output of the IC (see Figure 12.4) to increase the complexity of logic on the chip. For example, in the case of truly static shift registers, clock generation can be incorporated on the chip, so that not only are the inputs and outputs TTL compatible, but so is the single clock input.

12.3 APPLICATION

The benefits of MOS are particularly realized with shift registers and although it is economical to replace long delay lines, replacement of smaller blocks of storage using shift registers has not paid because of the high cost per bit of interfacing.

An example is a signal averager built by Data Labs, London, England. Averagers and correlators are designed to improve the signal-to-noise ratio of time-locked signals by accumulating a preset number of signal sweeps in a store and automatically computing the average result.

A basic system with some typical values is shown in Figure 12.6. Each signal sweep consists of 100 sampling points, at each point a 7-bit analog-to-digital conversion is performed and the result is arithmetically added to the value obtained at that same point on the previous sweep. Up to 512 (i.e., 2^9) sweeps are permitted, and consequently a 16-bit adder and store is required. The storage is provided by TTL-compatible dual 50-bit shift registers, the truly static characteristic of which is required at the slowest sweep speeds where the sample time is once every 0.2 second. Binary repre-

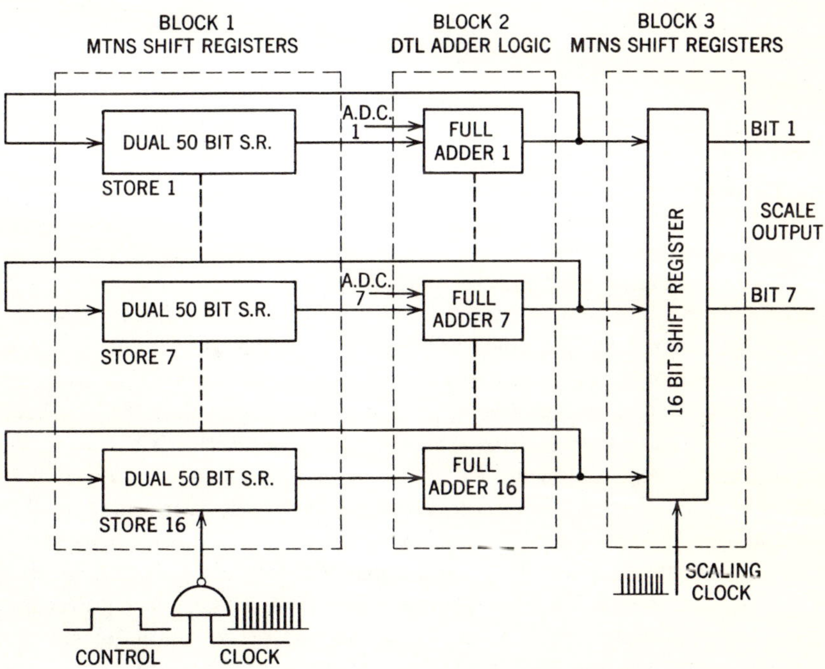

Figure 12.6. Application using shift registers as stores.

sentation is used throughout and it is a simple matter to take the average value by shifting the accumulated totals in yet another register.

The addressing of each bit is not required in this application; it is only necessary to synchronize with the first bit at the beginning of the sweep. Because the register is truly static, it can be stopped after each sweep while other steps are performed; and if a stream of pulses 100 bits long is used to shift the register contents, triggered at the start of each sweep, no further synchronization or addressing is required.

This stream of pulses from the DTL gate can drive the register-shift input directly because in the registers used (type SL6–2050) the clocks are internally generated and only a single, bipolar-level-voltage signal is required.

Although ferrite storage systems (core, disk, or drum) may prove more economical for storage above 20,000 bits, up to this size, shift registers can be used as alternatives to glass or magneto-restrictive delay lines, providing more flexible addressing at the same time. Figure 12.7 illustrates such a system. Assuming a 16-bit word, the storage is arranged as 8-word blocks, using the dual 64-bit register, and all the words in the blocks are addressed and sequenced in parallel, with the blocks being addressed sequentially. The word addressing is carried out continuously. As the contents of the register are shifted by the basic clock, the word-address counter increments by one every 16 clock bits to maintain synchronism. When concidence with the word address is obtained, the read/write lines are gated for the 16 pulses required to read or write the information at the selected block. Blocks are selected serially, each block having its own shift-register bit, one stage of an SS6-2004, and a dual 4-bit shift register. A single logical "1" is shifted from block to block until block-address coincidence is achieved, at which time further block-address shift pulses are inhibited.

There are several advantages to using MOS ICs. For example, more blocks can be added simply by breaking into the block-address shift-register loop and increasing the block-address counter size. Furthermore, access time is kept down because searching for a block and for a word is performed simultaneously; thus the longest access time for the system in Figure 12.7, using a 1-MHz clock, is 128 μs. As soon as a new address is set up at the input, it is automatically searched for. If a word address is detected first, it is held until block-address coincidence releases the word-address shift line and the 16 bits are read from or written into the store. Should the block address be detected first, the storage registers are accessed as soon as the word address is reached. This system is useful where low power and wide environmental conditions preclude the use of conventional storage. In situations where the speeds of standard shift registers are inadequate,

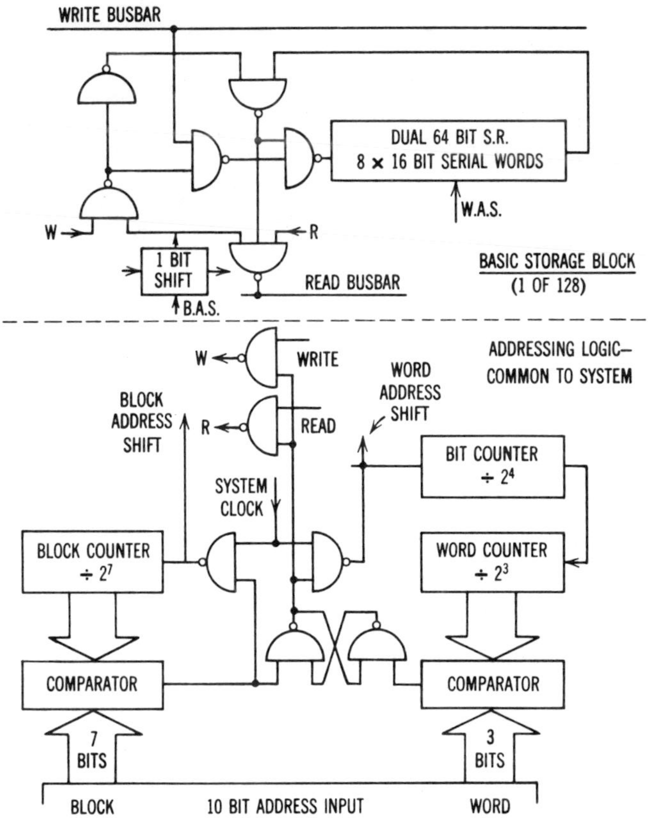

Figure 12.7. Addressing logic for shift-register bulk storage.

a time-sharing system in which the operations are data time encoding-shifting, data time decoding can be constructed.

Comparatively low-speed systems such as digital voltmeters lend themselves to MOS applications. A specific example is the digital panel meter. It is no problem with MOS packing densities to put the counter store and decode sections on the chip to provide a universal counting system. By using such an IC, only three identical packages are required, board space and power are saved, no extra power supplies are required since the +5 V of the logic and —12 V V_{DD} of the analog amplifiers are used, and no interfacing is required between the bipolar logic and the MOS systems.

13

Character Generation

by Richard B. Mann

Texas Instruments Incorporated

This chapter demonstrates the employment of the MOS digital storage buffer (DSB), the static shift register (SSR), and the read-only memory in designing a display system. The system is intended to be used with any standard oscilloscope and in its present form is capable of generating a row of 32 alpha-numeric characters on a cathode ray tube (CRT). Fresh characters can be typed in from a keyboard and these will be entered on the right-hand side of the display at a steady rate of two characters per second even if the typing speed exceeds this rate. The display, therefore, moves to the left in the manner of a newscaster billboard.

13.1 SYSTEM DESCRIPTION

A block diagram depicting a basic system for character-generation display is shown in Figure 13.1. The MOS ROM used contains enough data for generating 64 characters, therefore 6-bit character-address codes are required. These codes can be obtained from a Honeywell solid-state keyboard having six strobed parallel outputs whose levels are compatible with TTL or DTL.

The system to be described in this chapter uses ICs built with high-threshold-voltage devices. Therefore, the data input code is taken to an interface unit in order to translate the logic voltage levels of TTL (logical "1" = +2 V, logical "0" = 0.8 V) into MOS logic levels (logical "1" =

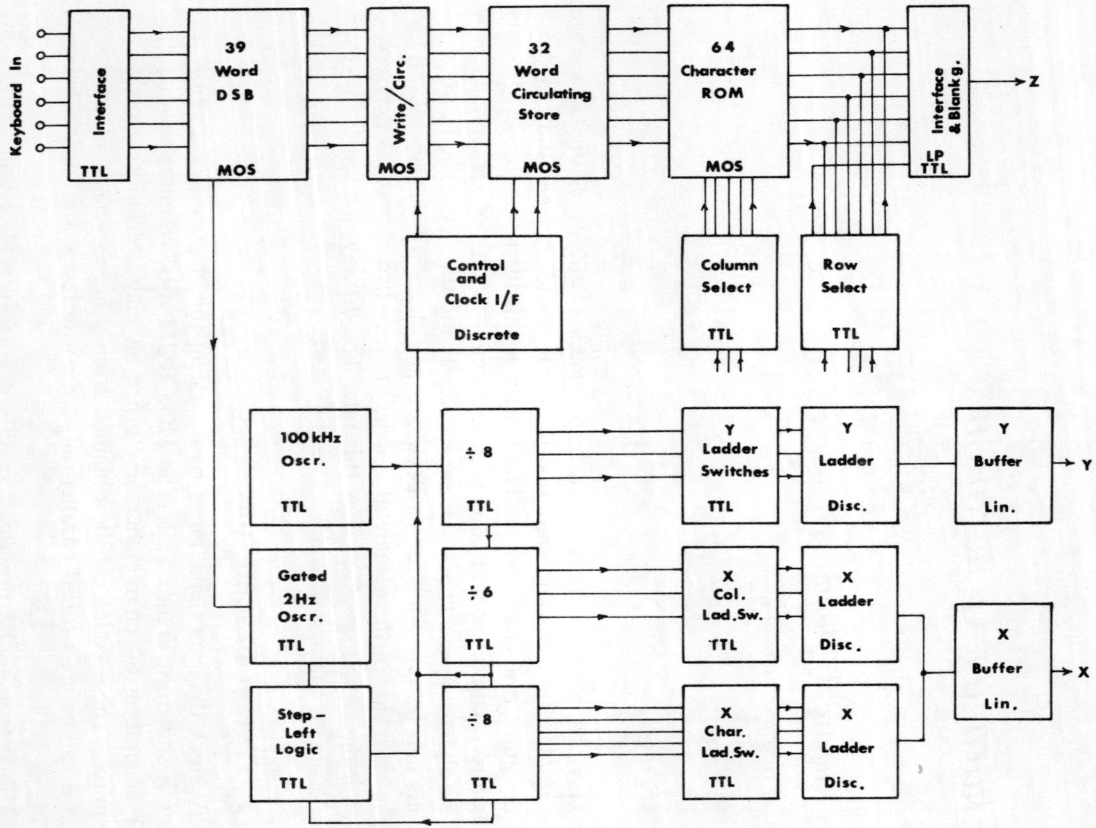

Figure 13.1. Basic character-generation display system.

—14 V, approximate logical "0" = 0 V). The character addresses are clocked at typing rate into a digital storage buffer capable of storing 39 words.

The addresses are then clocked out of the DSB at a steady rate of 2 Hz into a circulating store that uses three dual 32-bit static shift registers. Thus, thirty-two 6-bit character addresses can be continuously recirculated. The output of the circulating store is applied to the address inputs of the read-only memory so that the display of 32 characters is completely refreshed with every circulation of the store. Each character display is based on a 7×5 dot matrix, and the information contained in the ROM determines which dots are brightened and which dots are suppressed.

With a single-beam CRT, the modulation data must, of course, be fed to the tube in sequential form. The ROM used actually has seven outputs that make a complete column of data available in parallel form. However, these data are serialized by using a seven-way row selector and a five-way column selector to pick out any particular dot in the matrix. Thus, and individual bit is decoded in the following sequence: first, apply character address; second, apply column address; third, apply row address. The column and row selections are performed with standard TTL decoders driven from a master control section, also using TTL. The outputs of the ROM can be applied to a low-power TTL gate without any form of active interfacing. This makes the row selection very simple and also allows a blanking pulse conveniently to be inserted in order to suppress the CRT beam when it is moving between dot positions.

The master logic section basically comprises an oscillator and buffer, which clocks a standard divide-by-8 counter at 100 kHz. The outputs of this counter provide the seven row-select conditions, plus a further one giving a margin at the bottom of a character. The outputs also drive a simple digital-to-analog converter (DAC), used to provide the Y deflection to the CRT beam. The most significant output of the divide-by-8 counter then clocks the divide-by-6 counter, which operates in a similar way, giving five column addresses plus a margin and also driving a DAC, which provides an X deflection output.

The output of the divide-by-6 circuit, in turn, is used to clock a 5-bit binary counter, which drives a third DAC. This provides further X deflection and thus gives 32 character positions across the face of the CRT.

Further logic is added to give a "step-left" function, which is clocked by a 2-Hz generator. The output of this generator, however, is gated by the digital storage buffer and operates only when there is a new character address waiting to be applied to the circulating store. The shift registers used in the circulating store require a two-phase clock pulse having almost twice the voltage swing required for MOS data inputs. This is obtained from a simple discrete circuit that is driven by the most significant output of the divide-by-6 counter.

13.2 DIGITAL STORAGE BUFFER

A DSB acts as a "smoothing capacitor" between digital information arriving at random rate and a system that needs this information presented to it at a uniform rate. This section of the system uses three separate DSBs (such as type TMS2A-400IMH), each of which can hold thirteen 6-bit words. The operation of the devices is analogous to a silo: As new words are clocked in, they automatically move down until they come to an unoccupied space as near to the output as possible. They can then be clocked out by a data output clock (DOC) pulse at a steady rate, so long as the data input clock (DIC) is supplying new words, i.e., character addresses. Each DSB requires nominal power supplies of −14 V (V_{DD}) and −24 V (V_{GG}). The power dissipation is typically 250 mW and the DIC and DOC pulses can have a maximum repetition rate of 250 kHz. The DSBs can be connected in cascade or in parallel to extend system storage capacity in increments of six digits and/or thirteen characters. To extend the character positions, two controls are provided: a next-register-input clock (NRC), which provides the input clock signal to the succeeding register in a cascaded string and a previous-register-output clock (PRC), which provides the output clock signal to the preceding register in the string. There is no limit, in terms of performance, to the number of registers in a cascaded string. To extend the digit positions, the input and output control signals of a number of registers are simply tied in parallel.

A diagram of the DSB section of the system is illustrated in Figure 13.2. It shows the interconnection required for storing 39 six-bit words. Each of the DSBs has a number of other facilities. There are flag outputs which indicate that there are words in selected positions: the first (FL_1), tenth (FL_{10}), and thirteenth (FL_{13}). In addition, a linear "dipstick" output is provided, which produces an analog current proportional to the state of fill of the first ten register positions. There is also an output to show "any data in storage" (ADIS); this is used to indicate lock-up conditions in the register. If the ADIS output and the first flag do not agree, the register is in lock-up and must be cleared to restore normal operation. This may be done by taking the clear input to a logical "1" for less than 30 μs. Lock-up usually occurs only if either power supply is momentarily removed. The flag and ADIS outputs are current sources (free drains) and require a load to V_{SS} if used.

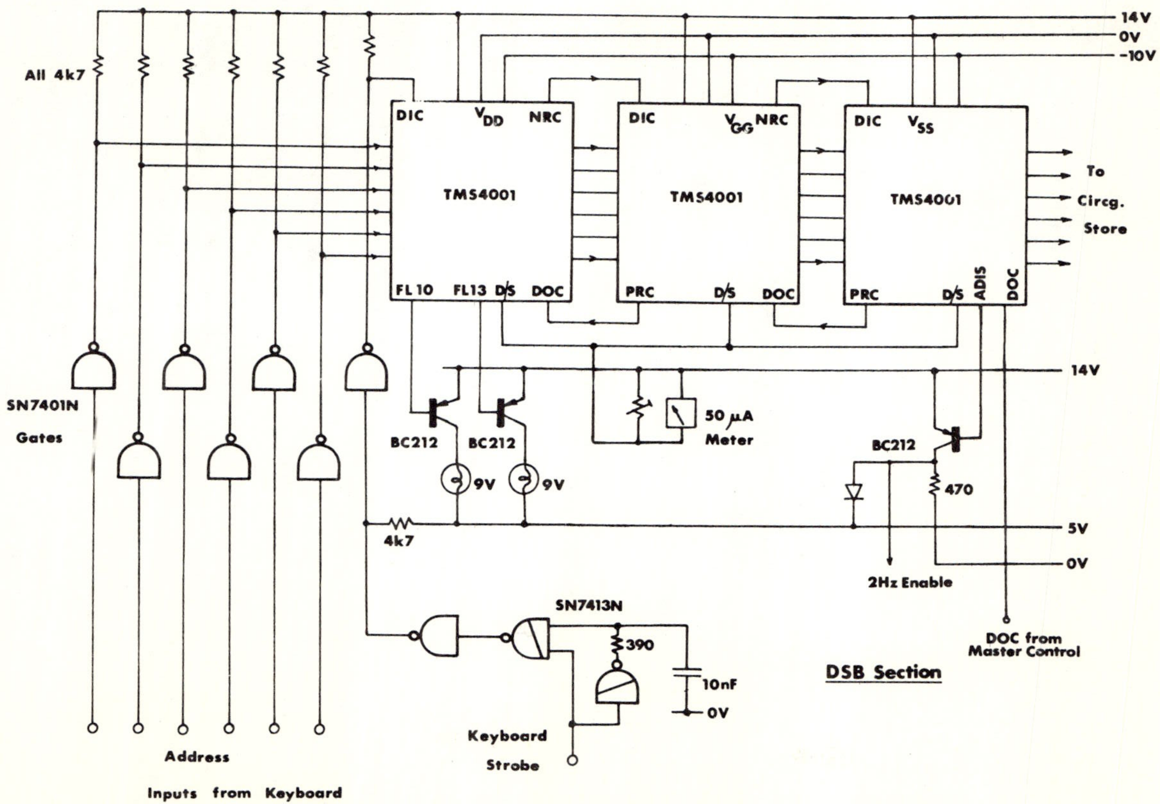

Figure 13.2. MOS digital storage buffer section of system.

13.3 DSB CLOCK INTERFACE

The DSBs, in common with other MOS devices, are connected with their substrates returned to a potential of +14 V. This is common practice when MOS and TTL are used in a system, since it allows interfacing without active level inversion. The data and DIC inputs are driven from two SN7401N quad two-input NAND gates with open collector outputs having greater than 14-V breakdown. When the output of the gate is at a logical "0" (approximately 0 V), the MOS input will see a logical "1" (−14 V to substrate). When the eight inputs are at logical "1," the MOS inputs will be pulled up to substrate potential and the MOS will see a logical "0." When clocking in data, it is essential that logical "1" data inputs overlap the DIC on leading and trailing edges. If a "0" appears at the input during any part of the DIC, a "0" will be loaded.

A simple differentiator circuit is included to produce a narrow DIC pulse from the keyboard strobe pulse. The differentiator uses and SN7413N, which is a TTL device logically and physically interchangeable with the SN7420N dual four-input NAND gate; but each of the gates in the package has Schmitt action with an accurately maintained hysteresis of typically 0.8 V.

Because of the regenerative action of the Schmitt gate it is possible to drive it satisfactorily from very slowly rising edges. This is unlike normal TTL, which requires edges in the order of 30 ns; thus a delaying capacitor (capacitor *C* in Figure 13.3) can be used without the danger of threshold oscillation at the input of gate G_2. This arrangement allows a narrow negative-going pulse to be derived from a wide positive-going pulse. The SN7413N also can be used as a very simple and stable clock pulse generator.

The clock interconnections of the three DSBs are shown in Figure 13.2. The NRC outputs can be connected directly to DIC inputs. No extra gating

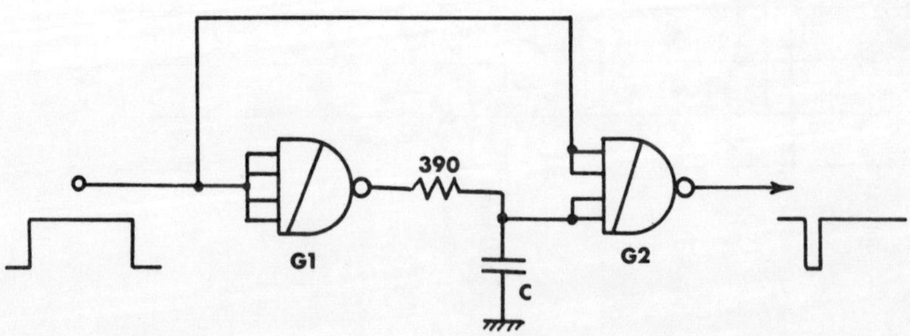

Figure 13.3. Use of delaying capacitor with Schmitt gate.

is required, since the devices will automatically provide an NRC pulse only if there is room in the succeeding DSB; furthermore, they will provide a PRC pulse only if there are data waiting in the preceding DSB. Since the dipstick outputs are current sources, they may be connected in parallel and the combined output may be fed to a 50-μA meter. A shunt resistor is provided across the meter to give a full-scale deflection when the dipstick output is at a maximum, i.e., 36 words in store. In a practical system, dipstick outputs can alternatively be used to control the DOC rate so that this keeps pace with the DIC frequency.

The FL_{10} and FL_{13} outputs on the first DSB drive the bases of two *pnp* transistors connected between the +5 and +14-V rails, which act as buffers driving small 9-W lamps. Depending on which one is on, these lamps indicate when the total store contains 36 words and 39 words, respectively.

The ADIS output on the third DSB gates the 2-Hz oscillator so that the display will shift left only when there is a new character waiting in the storage buffer.

Since the circulating store can accept new information only at certain times during its cycle, the output of the digital storage buffer must be synchronized with the write-circulate function of the circulating store. This synchronizing pulse is obtained from the master control section via a further interface circuit. The system design is simplified if the DSB devices described are replaced by low-threshold DSBs.

13.4 THE CIRCULATING STORE

A circulating store section using three dual, 32-bit, static shift registers such as the TMS7B3001 will meet the system requirements. The SSRs are connected in parallel to give a capacity of thirty-two 6-bit words and are clocked at approximately 2.1 kHz (i.e., 100 kHz divided by 48); consequently, they will shift through 32 character addresses in 15.4 ms and refresh a display approximately 65 times per second.

The outputs of the SSRs are returned to their inputs via MOS NOR gates (such as type TMS1A1702LA). The gates are arranged as shown in Figure 13.4 to provide a two-way switch at the input of the shift register. Thus, when G_1 is enabled and G_2 is inhibited, data will constantly recirculate in the SSR; and when the opposite situation occurs, new information can be written into the shift register via gate G_2. Each NOR gate has three inputs; the spare input therefore, mut be returned to an MOS logical "0." The right and circulate control lines must always be driven in antiphase and, therefore, are derived from Q and $\overline{Q}$ outputs of a flip-flop in the master control section. The SN7401N open collector gates are again used to provide

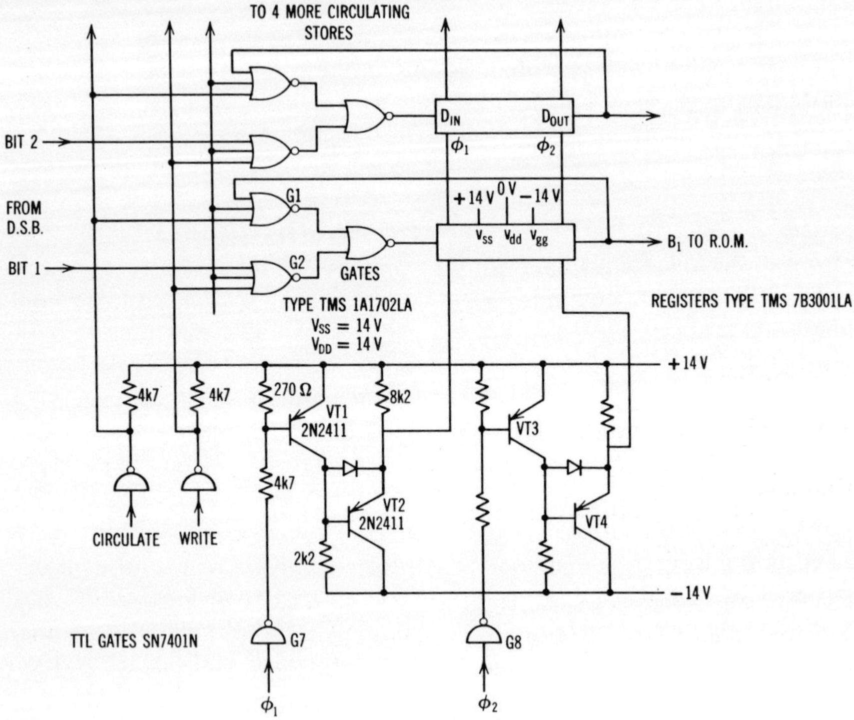

Figure 13.4. Circulating store section.

the TTL-to-MOS interfaces. The two remaining gates in the package are used to drive the two-phase clock generator.

Figure 13.5 shows a single stage of the shift-register (transistors VT_1 to VT_7) and the gating transistors, VT_8 and VT_9, which are used to generate the third clock ϕ_3. Figure 13.6 defines the clock waveforms that are necessary to perform a single shift operation. For the clock pulses a logical "0" is nominally 0 V and a logical "1" is nominally −28 V with respect to substrate. For static data storage it is essential that ϕ_2 be held at "1" while ϕ_1 is held at "0." The shift stage comprises a cross-coupled latch with a master–slave action. Transistors VT_3 and VT_6 serve merely as drain resistors for the active devices VT_2 and VT_5. When it is required to shift or enter information, ϕ_2 is initially returned to logical "0," which turns off transistor VT_4. The charge held on C_3 decays and transistor VT_5 is turned off, so that latching action is disabled. After a minimum period of 10 ns, ϕ_1 is taken to a logical "1," which turns on VT_8, clamping ϕ_3 to "0" and also turning on VT_1, which allows data to be entered. Capacitors C_1 and C_2 are capable of maintaining the logic levels for at least 10 μs in the absence of the latch. This sets the maximum periods as indicated in Figure 13.6.

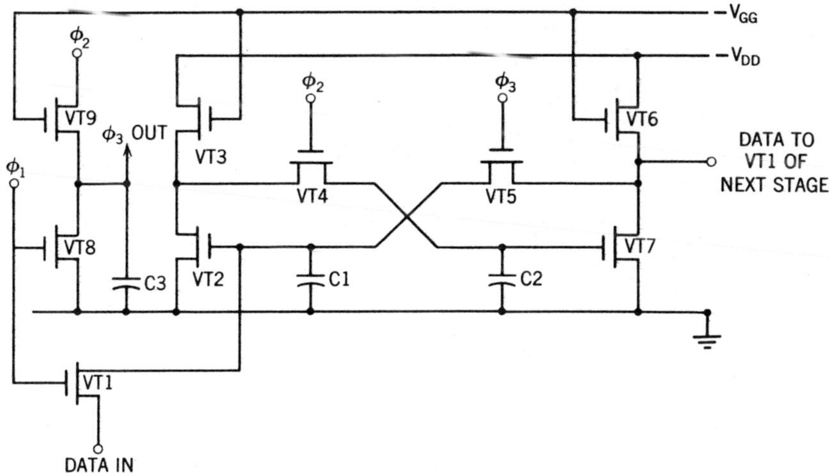

Figure 13.5. Clock generator and single stage of static shift register; transistors VT_8 and VT_9 are used to generate the third clock.

Since VT_5 is off, the data input sees a purely capacitive load of approximately 2 pF. Now ϕ_1 can be returned to "0," turning off transistors VT_1 and VT_8 so that data are locked out. The new data now appear at the drain of transistor VT_2, with the previous data still held at the drain of VT_7. After

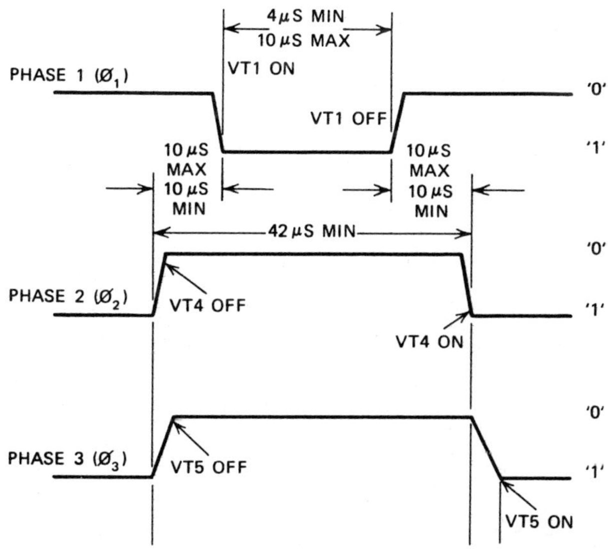

Figure 13.6. Clock waveforms for a single-shift operation; maximum rise and fall time for ϕ_1 and ϕ_2, 5 μs.

a further minimum period of 10 ns, ϕ_2 reverts to logical "1," which turns on transistor VT_4 and transfers the new data on to capacitor C_2 and hence to the drain of transistor VT_7.

It is important to remember that ϕ_3, the gate drive to transistor VT_5, is delayed by the time constant associated with capacitor C_3 and the on resistance of transistor VT_9, so that the new information can transfer into capacitor C_2 without being erased from capacitor C_1 by the previous information at the data output.

Since clock shift registers represent an almost entirely capacitive load and require a 28-V drive at frequencies up to 1 MHz, the two-phase clock generator requires fairly careful design if a number of them are to be driven in parallel.

The circuit shown in Figure 13.4, the simplest of a number of possible circuits, is capable of driving capacitive loads up to 500 pF at frequencies in excess of 1 MHz. The only restriction on the antiphase input pulses is that they be of less than 10-μs duration if register operation at 80°C is required. Gates G_7 and G_8 act as buffers and level translators driving the bases of VT_1 and VT_3, respectively.

When the TTL input level to G_7 changes from a logical "0" to a logical "1," transistor VT_1 is driven into saturation and the base drive is removed from transistor VT_3. The collector VT_1 rapidly rises to $+14$ V, charging the ϕ_2 load through diode D_1. Because of charge storage, transistor VT_3 does not turn off until about 200 ns after its base drive has been removed. The ϕ_1 load capacitance is then discharged through transistor VT_4, rapidly taking ϕ_1 to -14 V. The same circuit operation occurs when the input to G_7 returns to a logical "0." Since the input to G_8 is in antiphase, transistor VT_3 is now rapidly driven into saturation and, after a short delay owing to storage, transistor VT_1 comes out of saturation. The resultant waveforms are shown in Figure 13.7.

Although it is not required in the system being discussed, it is worth mentioning that there are several methods of clearing or presetting the MOS static shift registers described above. Probably the simplest of these methods is to take both clock input phases to a logical "1" (i.e., negative with respect to substrate) and take the serial input to a logical "0" or a "1," as required. The level applied to the serial input will then ripple through all the stages in the shift register within a few microseconds.

13.5 READ-ONLY MEMORY SECTION

The read-only memory device used is a TMS2A4842MH. The memory matrix of the 484*X* series of ROMs comprises 2240 *p*-channel MOS devices arranged in a 32×70 array. MOS FETs are arranged with their

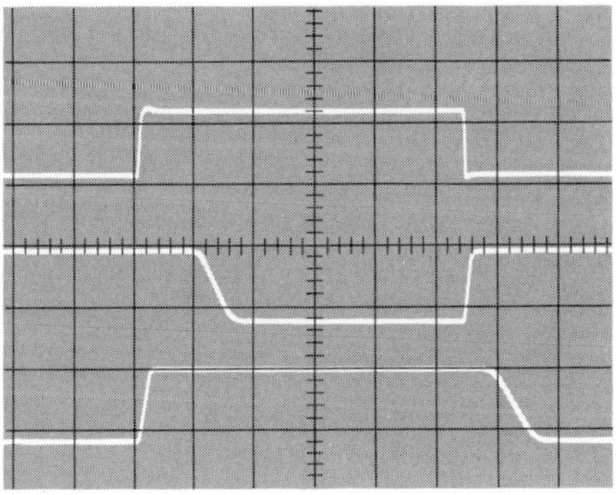

Figure 13.7. Waveforms for circuit shown in Figure 13.4. Input (top waveform) coordinates: 5 V/cm, 200 ns/cm; output ϕ_1 (middle waveform) coordinates: 20 V/cm, 200 ns/cm; output ϕ_2 (bottom waveform) coordinates: 20 V/cm, 200 ns/cm.

gates driven by 32 horizontal lines, with each drain connected to one of 70 output lines. All the device sources are connected together. When one of the 32 gate lines is driven negative, the devices that have their gates connected are turned on, pulling down the voltage of the corresponding output line. The mask programming determines whether the gate is connected; when a gate is connected, a dot is produced in the output format. Many ROMs are programmed at the second oxide-removal processing stage so that unused devices have a layer of thick oxide over their gate regions. When a logical "1" is applied to one of the five column-select inputs, seven of the 70 drain lines are connected through buffers to the output terminals.

With the 64-character ROM, two characters are included on each of the 32 gate lines. Full character-address decoding is included on the chip, therefore five of the six character-address bits are used to drive the address decoder and the sixth bit is connected to the output buffer to choose between the two characters.

The power supplies required are nominally 0, —14, and —24 V for substrate, V_{DD}, and V_{GG}, respectively. No current is drawn from the V_{GG} supply since it is connected only to MOS gates. The drain current I_{DD} is approximately 25 mA, resulting in 350-mW power dissipation. As before, the substrate is actually connected to a potential of + 14 V, V_{DD} is 0 V and V_{GG}, in this case, is —10 V. Therefore 0 V at any address input represents a logical "1" and +14 V represents a logical "0." Since the 6-bit character address code is obtained directly from the MOS circulating store, no interface circuitry is required. The five column-select lines, however, do require an interface with the TTL circuitry and an SN7445N TTL decoder (Figure 13.8) is used. The SN7445N is a four- to ten-line binary decoder having open collector output devices that can be returned to a +14-V rail.

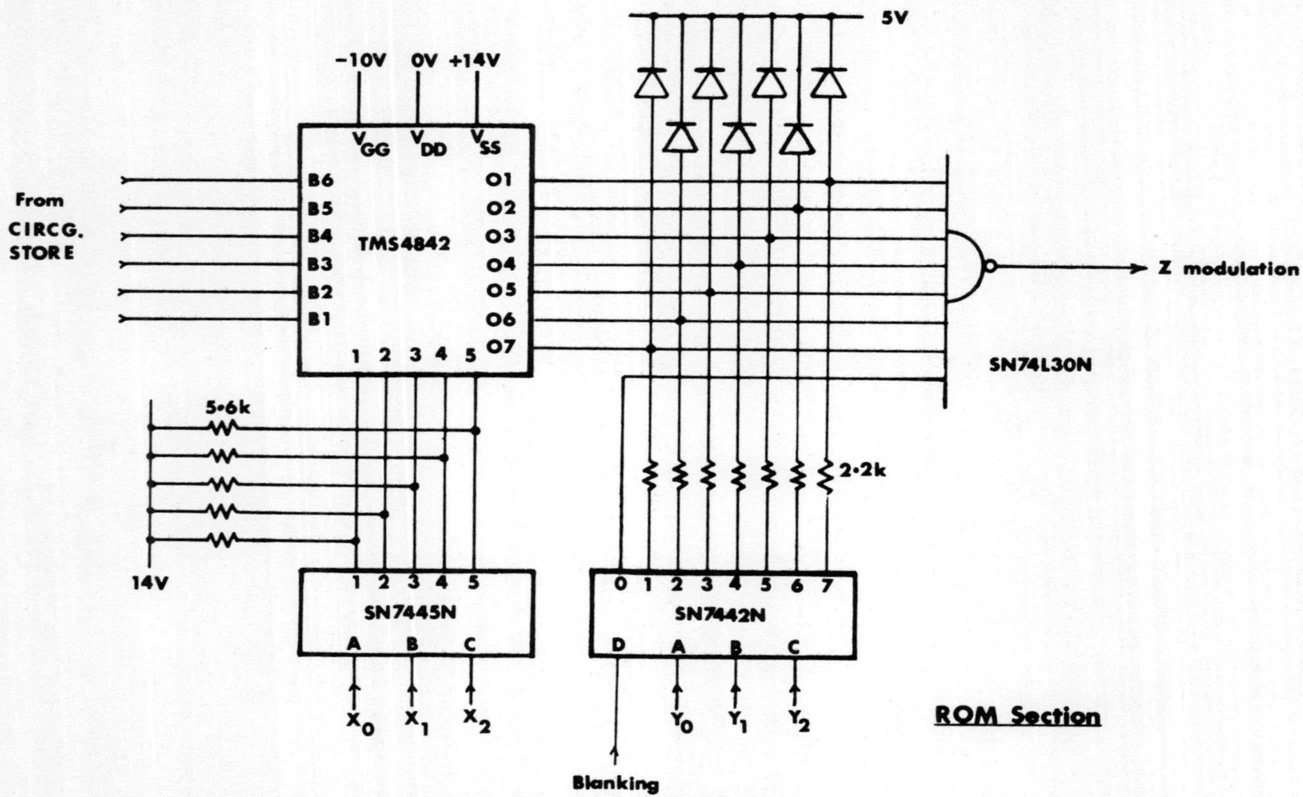

Figure 13.8. Read-only memory section.

Only five of the ten output lines are required, and these are derived from the binary address code X_0, X_1, X_2.

Output interfacing and row selection from the seven parallel outputs are performed very easily by means of a low-power TTL gate (type SN74L30N). This is an 8-input NAND gate and, since low-power TTL runs at currents approximately one order of magnitude lower than standard TTL, the maximum input current of the gate is only $-180\ \mu A$.

The minimum current obtained from a ROM output which is on (i.e., dot enabled) is 300 μA. This current effectively flows from the +14-V rail via a current generator. It is, therefore, possible to choose a value of output load resistor for the ROM that will ensure a full TTL logic swing at the input of the low-power gate. A suitable value is 2.2 kΩ. Since more than 300 μA may be supplied by the current generator, it is advisable to use diodes to clamp the gate inputs to the +5-V rail in order to avoid the possibility of emitter-to-emitter breakdown in the gate. Row selection is performed by returning the 2.2-kΩ resistors to the outputs of an SN7442N decoder similar to the SN7445N, but with normal TTL totem-pole outputs. When a particular output is selected, its level falls to a logical "0," leaving the other outputs at a logical "1."

This is very convenient since it means that the low-power gate inputs associated with ROM outputs that are not selected will be forced up to a logical "1" by the decoder. Therefore, the gate will be enabled and its output will be determined solely by the ROM output selected. Only seven of the ten decoder outputs are used for row selection, but an eighth output, corresponding to a binary input code 0, 0, 0 is used to inhibit the low-power gate. This means that the CRT beam is suppressed during the column flyback time and must return to margin position, thus giving the Y deflection DAC more time to settle.

13.6 MASTER CONTROL SECTION

This section consists entirely of TTL circuitry as shown in Figure 13.9. The first part is concerned with producing the thirty-two 7×5 dot matrices and the second part with the "step-left" function and the entry of new information from the DSB. The first part is clocked by a simple relaxation oscillator based on S_1 (half an SN7413N Schmitt trigger package). The lower threshold level of these devices is nominally 0.9 V and the upper threshold level is 1.7 V, and as mentioned above, hysteresis is accurately maintained. This means that the input will alternate between the two threshold levels with the capacitor C (in Figure 13.9) being charged and discharged alternately via the output of S_1 and the 390-Ω resistor R_1.

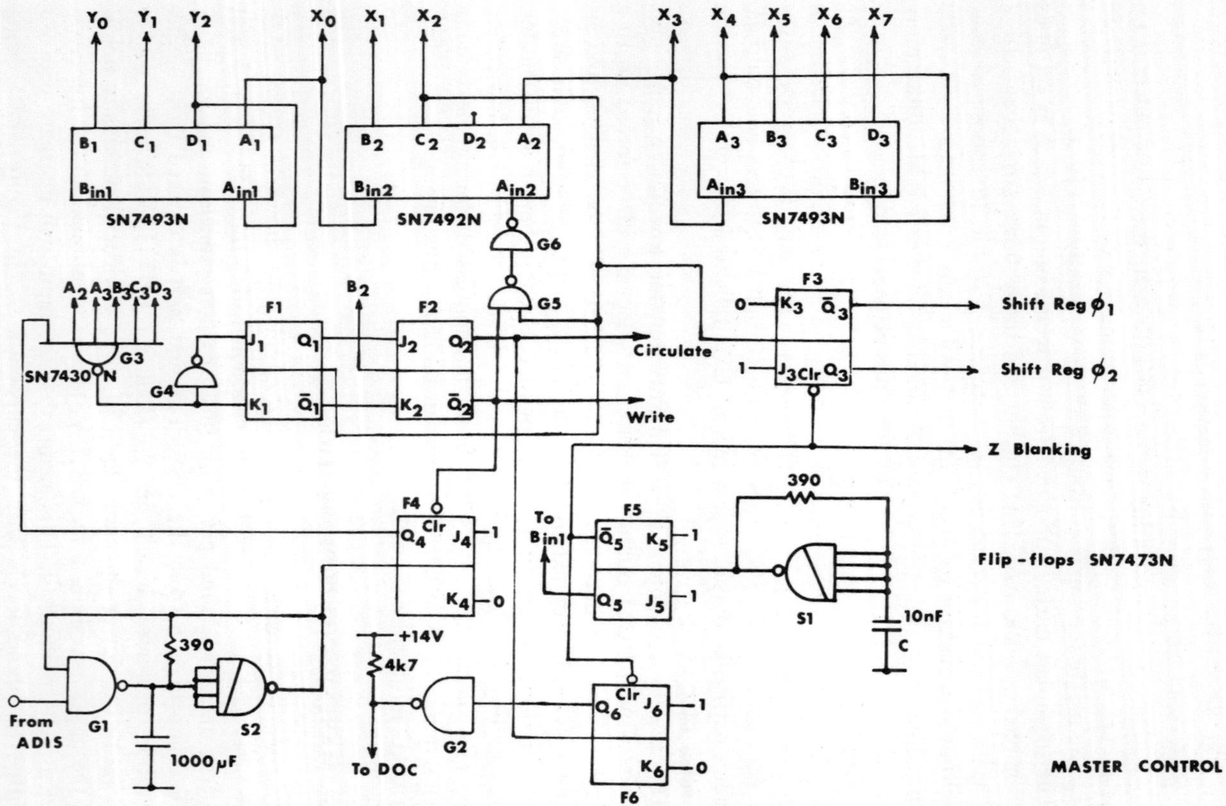

Figure 13.9. Master control section consists entirely of TTL circuitry.

The time constant is chosen to give a pulse repetition frequency of 200 kHz, and the output of S_1 is also used to clock a J–K flip-flop F_5, which acts as a buffer. Output Q_5 drives an SN7493N 4-bit binary counter. This device is arranged as a 3-bit counter having outputs B_1, C_1, and D_1 and a separate 1-bit counter A_1. The 3-bit counter supplies the row-selection code Y_0, Y_1, Y_2 to the row decoder and also drives the Y deflection DAC. The remaining binary section in this package is combined with an SN7492N counter. The latter is also split into two parts comprising a modulo-six counter and a separate binary stage. To ensure that the modulo-six output has the binary weighting required for the address X_0, X_1, X_2, it is necessary to precede the divide-by-6 section by a binary stage. Therefore X_0, X_1, and X_2 are derived from A_1, A_2, and C_2 rather than from B_2, C_2, and D_2.

The ROM column selection is obtained from the address X_0, X_1, X_2, and this code also drives the three less-significant bits of an 8-bit DAC. The output of the remaining binary stage in the SN7492N package (A_2) is used with a second SN7493N 4-bit binary counter, thus providing the inputs to the five more-significant bits in the X deflection DAC. This gives the 32 character positions.

When all 35 dots have been addressed in a particular character, the X_2 address line falls from a logical "1" to the "0" level. This negative edge is used to clock the J–K flip-flop F_3, giving a logical "1" at the ϕ_2 input to the shift-register interface. Since the SN7492N and SN7493N counters trigger on a negative-going edge at output Q_5 the ϕ_2 shift pulse will be synchronized to this edge. Flip-flop F_3 is cleared by a "0" at its clear input and this is provided by output Q_5, whose level falls 10 μs later. The two-phase clock pulse to the static shift registers is therefore within the maximum limit specified.

The entry of new characters and the step-left action are timed by the gated 2-Hz oscillator comprising S_2 and G_1 in Figure 13.9. Gate G_1 is an open collector NAND gate which, when it is enabled, clamps the input of the Schmitt gate to earth and inhibits the oscillator. Feedback from the output of S_2 to the input of G_1 ensures that the oscillator always completes its full negative-going pulse width. A logical "0" is required on the second input of G_1 to enable the oscillator, and this is provided by the ADIS output of the final DSB via an interface transistor.

When a clock pulse is received from the oscillator, the Q_4 output of flip-flop F_4 rises to a logical "1," allowing the 8-input gate G_3 to be enabled when X_3 through X_7 rise to a logical "1" (that is, when the CRT beam is in its extreme right-hand position). The subsequent operation of the "step-left" action is shown in Figure 13.10. After the thirty-first shift-register clock pulse, the circulating store has character 32 at its output and the CRT beam is in position 32 as indicated in Figure 13.10*a*. On the

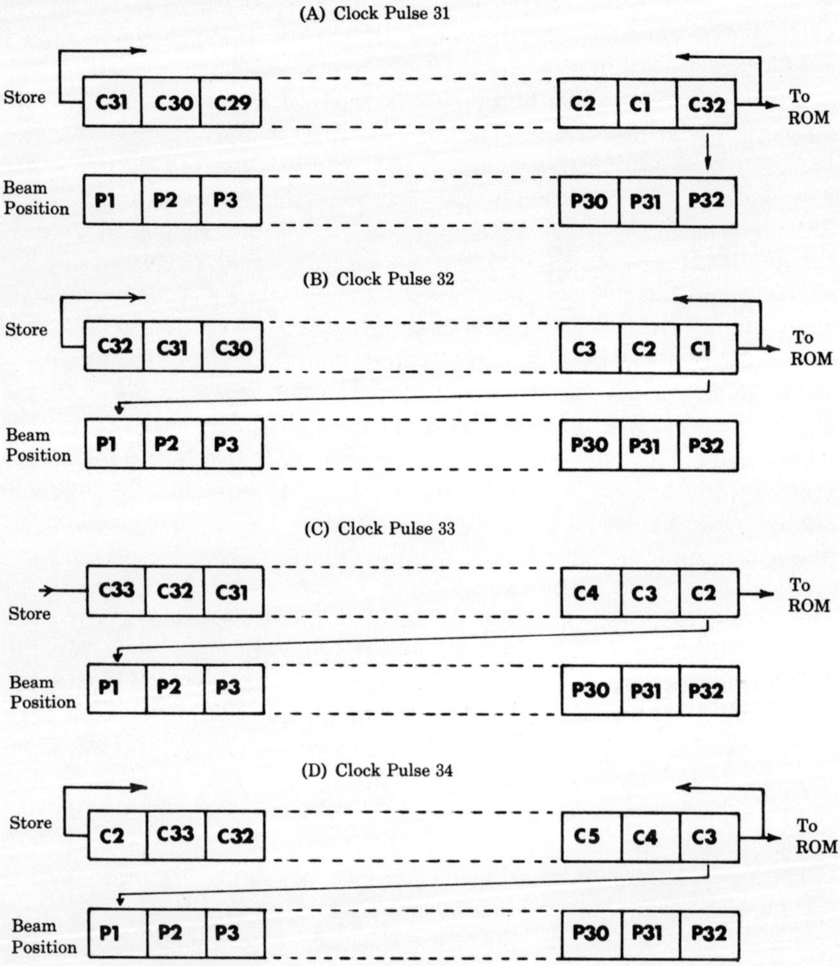

Figure 13.10. Operation for clock pulses 31 through 34.

thirty-second shift-register clock pulse, character 1 moves to the store output and the beam returns to position 1 as indicated in Figure 13.10*b*. At the same time, the output of G_3 is put into flip-flop F_1, since the propagation delays in the circuits ensure ample hold time at inputs J_1 and K_1. While character 1 is being serialized, a negative edge appears at counter output B_2, which then shifts the output of F_1 into F_2.

The output of F_2 resets F_4 so that it will be ready to receive another clock pulse from the 2-Hz oscillator. At the same time the counter input A_{IN_2} is inhibited by gate G_5 so that the CRT beam will be held back in character position 1. The write line of the circulating store is enabled and the circulate line is inhibited, and finally the DOC pulse is supplied to the DSB via flip-flop F_6 and interface gate G_2.

On the thirty-first shift-register clock pulse, character 2 arrives at the output of the store and is displayed in position 1 since the position counter has been held back. This is illustrated in Figure 13.10*c*. Simultaneously a new character, 33, is written into the store instead of character 1, which is lost. As before, the negative edge appears at counter output B_2 prior to the next shift-register clock pulse, which resets F_2, thereby enabling the character position counter and restoring the store to its circulate condition. Thus on the thirty-fourth clock pulse, the displayed message will have shifted one space to the left as shown in Figure 13.10*d*.

All CRT beam movements are synchronized to a negative edge at output Q_5. In order to reduce smearing, the beam is cut off while it is moving by taking output $\overline{Q}_5$ to the D input of the SN7442N decoder in the ROM section. This is equivalent to a 2^3 input, which means that all the decoder outputs 0 to 7 must be inhibited, thereby inhibiting the low-power interface gate.

13.7 DIGITAL-TO-ANALOG CONVERTER

The final section of the system comprises three DACs: one to provide Y deflection corresponding to row position, one to provide X deflection corresponding to column position, and the third to provide X-deflection corresponding to character position. The setup is shown in Figure 13.11. Each DAC uses a binary weighted ladder with very simple voltage switches. Considering the least significant bit of the Y DAC (Figure 13.11), the NAND gates have open collector outputs (SN7401N) and all the resistors are of equal value ($R\ \Omega$). Thus when the output of gate G_1 is at a logical "0," the junction of R_2 and R_6 is grounded via resistors R_3 and R_4 in series, since diode D is reverse biased. Therefore the correct ladder ratios are maintained.

The logical "0" output voltage of an SN7401N gate is typically 220 mV at a current of 16 mA, and variations in this voltage are insignificant compared with 14 V, since absolute accuracy is not required in this application. When the output of gate G_1 is at a logical "1," it is effectively an open circuit. Current flows from the +14-V reference rail through the diode, bypasses R_4, and continues to give the correct resistance ratio. The output current of the ladder flows into the virtual ground provided by operational amplifier OA_1, which acts as a buffer.

The X deflection DACs are very similar to the Y DAC, and the outputs

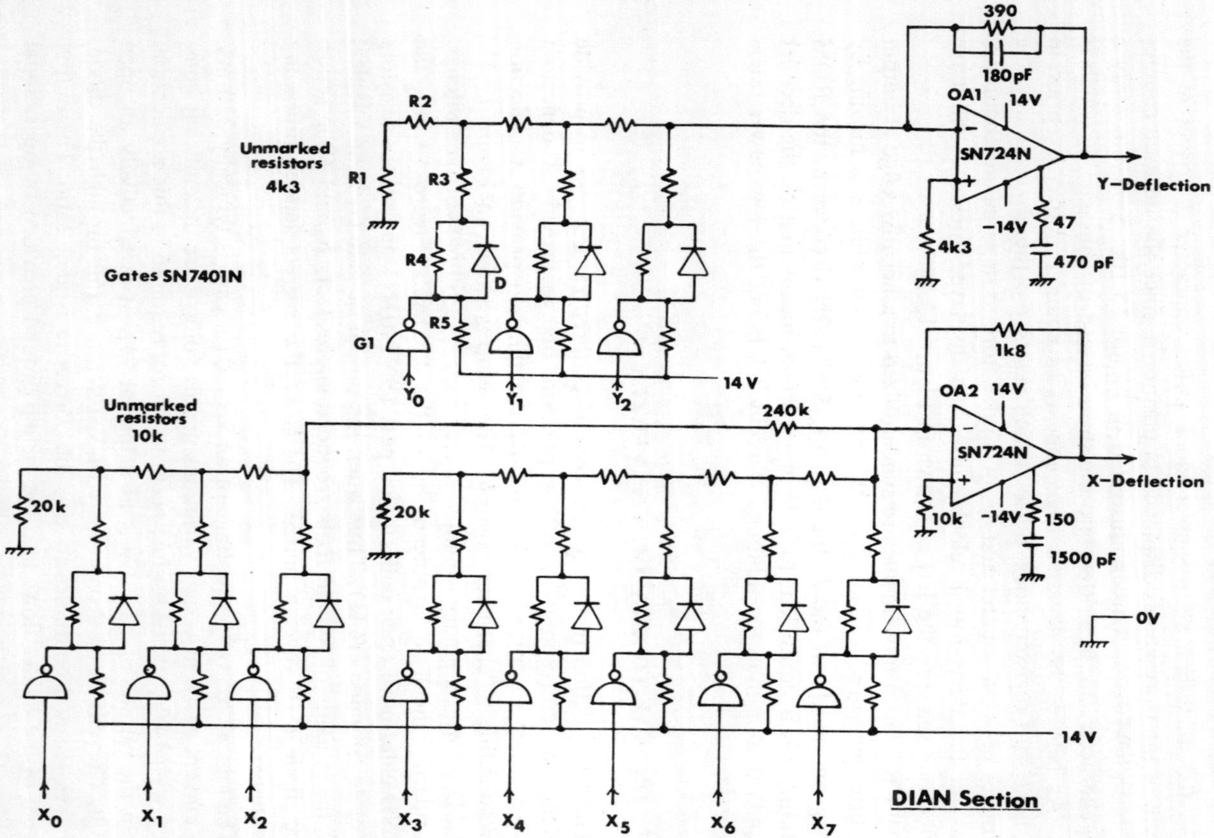

Figure 13.11. Digital-to-analog converter section.

of the two ladders are combined in a current summing buffer amplifier OA_2. The output of the 3-bit X ladder is weighted to give a maximum output equal to one thirty-second of the maximum output from the 5-bit X ladder. The gains of the X and Y buffer amplifiers are adjusted to each other so that the correct format is obtained when the X and Y sensitivities of the CRT display unit are both set to 200 mV per centimeter.

13.8 CONCLUSIONS

The system described was designed as an exercise in combining MOS and bipolar integrated circuits with minimum complication. The system was built and nicknamed the Message Maker. Obviously the system can be simplified if low-threshold MOS devices are used to perform needed circuit functions.

There are other ways in which the system might be improved. For example, the MOS NOR gates used could be eliminated by using a dual 128-bit static accumulator; this would considerably reduce the package count if used in the circulating store section, quadrupling the number of characters that can be displayed, as well. This performance would still be well within the speed capabilities of the read-only memory, which typically works at rates up to 1200 characters per second.

There are, of course, other ways in which MOS read-only memories may be used for character generation, making more efficient use of their parallel output facility. For example, the column-output ROM can be used with a large matrix of light-emitting diodes, which can be strobed a column at a time, thus avoiding the need for serialization. Similarly, a row-output ROM is available which is suitable for TV raster displays. Another possible application is the use of a single 4096-bit ROM as vector character generator. Here the ROM would be used to store the terminal coordinates of vectors that form a character, rather than following the procedure of storing dot information on a regular pattern. This should be an inherently more efficient system, since characters can take up more or less storage space according to their complexity.

ACKNOWLEDGMENT

The author acknowledges the assistance of Gordon Hoffman, Bob Parsons, and Derek Skinner (all of Texas Instruments) in the preparation of this material.

14

MOS Integrated Circuits in Microprogram Applications

by E. Floyd Kvamme, B.S.E.E., M.S.E.E.

National Semiconductor Corporation

Using today's MOS integrated circuits of both the serial-shift-register and read-only-memory configurations, it is possible to devise systems using microprogramming techniques. In these systems, the registers operate as serial data storage, whereas the read-only memories operate as the machine's microprogrammer. In the more general application of this technique, a system can be designed that uses the MOS read-only memories in a microprogramming capacity but substitutes TTL logic circuits for the parallel organized processor.

14.1 REGISTERS

Serial shift registers of both the dynamic and static varieties are commonly used as memory media in applications involving the replacement of a delay line or a small drum memory. A system advantage of using an MOS shift register is that it can be stopped for resynchronization purposes at any desired location while a particular operation is being performed. In a sequential system, this means that the average access time to serial memory is much less than the typical access time of 50% of the length of the delay line. Looking specifically at small machines, the MOS serial shift register has been used to perform the memory function of the accumulator registers. For example, it is now possible to place all data registers on a single MOS chip. This MOS device accomplishes the simple storage functions of the serial machine.

The operation of a typical arithmetic configuration is shown in Figure 14.1. In this case, the BCD bit serial added requires four clock times to accommodate a true BCD function. Information in the registers is stored in true BCDs to accommodate easier display at the device output. The function of the input gate structures is to steer the data from the adder into the appropriate storage register. The logic used in controlling the gate structures can be made up of other MOS devices but is most commonly constructed of TTL or DTL elements in simple machines.

14.2 READ-ONLY MEMORIES

Read-only memories are available that store 2048 bits of information. A typical configuration of such a ROM is shown in Figure 14.2.

The read-only memory has most commonly been applied where code conversion is being performed. For example, if the input word to a read-only memory were the value of an angle, the output word would be the

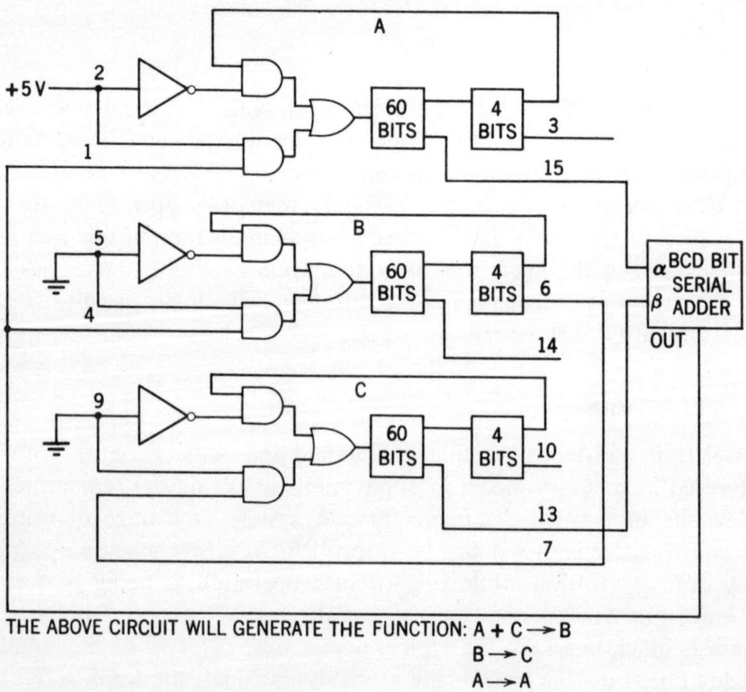

Figure 14.1. Typical arithmetic configuration using BCD bit serial adder.

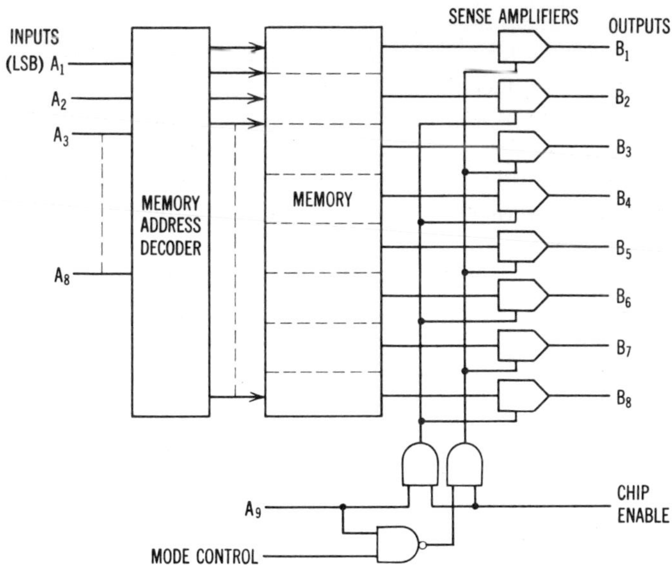

Figure 14.2. Typical configuration for a read-only memory.

sine of that angle. Similarly, the input word might be expressed in ASCII code, whereas the output code might be in IBM Selectric language. In these simple read-only memory applications, the unit is being used primarily to convert one word structure to an output word structure; that is, the read-only memory is being viewed as a device that creates an output word from an input word. This application is shown in Figure 14.3, where the table look-up or the code-conversion type of use is demonstrated.

An alternate but similar application of the read-only memory is made

Input word							Output word							
A_1	A_2	A_3	A_4	A_5	A_6	A_7	X_1	X_2	X_3	X_4	X_5	X_6	X_7	X_8
0	0	0	0	0	0	0	1	1	1	1	1	1	1	1
0	0	0	0	0	0	1	0	1	0	0	1	0	0	1
0	0	0	0	0	1	0	0	1	0	1	1	1	0	0
0	0	0	0	0	1	1	0	0	0	1	0	0	0	0
0	0	0	0	1	0	0	1	0	1	1	0	0	1	1
0	0	0	0	1	0	1	1	0	1	1	0	1	0	0
0	0	0	0	1	1	0	0	0	0	0	1	1	0	1
0	0	0	0	1	1	1	1	0	0	0	0	0	1	0
0	0	0	1	0	0	0	0	1	1	0	1	0	0	0
0	0	0	1	0	0	1	0	0	0	1	0	0	1	0

etc.

Figure 14.3. Conversion of one word structure to an output word structure.

in character generation in which the ROM input address lines are divided to form both a character address and a line or column address. The output character type of application is shown in Figure 14.4.

A more powerful application of read-only memories, in which the output lines are thought of more in terms of creating a logic truth table, is shown in Figure 14.5. Notice that the inputs are called input variables and the outputs are termed independent output functions, capable of being expressed in the six input variables. This technique has many varied applications in a control system. For example, in a very simple device having a limited number of input variables, it is possible to perform all logic equations within the read-only memory by simply programming the minimum sum function on the ROM program sheet. In this way, all logic

Character address				Line address			Output character							
A_1	A_2	A_3	A_4	B_1	B_2	B_3	f_1	f_2	f_3	f_4	f_5	f_6	f_7	f_8
0	0	0	0	0	0	0	1	0	0	0	0	0	0	1
0	0	0	0	0	0	1	1	1	0	0	0	0	0	1
0	0	0	0	0	1	0	1	0	1	0	0	0	0	1
0	0	0	0	0	1	1	1	0	0	1	0	0	0	1
0	0	0	0	1	0	0	1	0	0	0	1	0	0	1
0	0	0	0	1	0	1	1	0	0	0	0	1	0	1
0	0	0	0	1	1	0	1	0	0	0	0	0	1	1
0	0	0	0	1	1	1	1	0	0	0	0	0	0	1
0	0	0	1	0	0	0	1	1	1	1	1	1	1	1
0	0	0	1	0	0	1	1	0	0	0	0	0	0	0

etc.

Figure 14.4. Conversion to character generation.

Minterm Function Realization

Input variables							Output functions							
A	B	C	D	E	F	G	f_1	f_2	f_3	f_4	f_5	f_6	f_7	f_8
0	0	0	0	0	0	0	1	1	0	1	0	0	0	1
0	0	0	0	0	0	1	1	1	0	0	1	0	0	1
0	0	0	0	0	1	0	1	1	0	1	1	1	0	0
0	0	0	0	0	1	1	1	0	0	1	0	0	0	0
0	0	0	0	1	0	0	1	0	1	1	0	0	1	1
0	0	0	0	1	0	1	1	0	1	1	0	1	0	0
0	0	0	0	1	1	0	1	0	0	0	1	1	0	1
0	0	0	0	1	1	1	1	0	0	0	0	0	1	0
0	0	0	1	0	0	0	1	1	1	0	1	0	0	0
0	0	0	1	0	0	1	1	0	0	1	0	0	1	0

etc.

Figure 14.5. Logic truth table.

for this simple system is performed by directly implementing the truth table for the system. It is, of course, unlikely that such simple requirements will be sufficient. In more complex cases, the input address (previously labeled the input variables) can, in fact, be the operation code that has been received from memory and is intended to control the data flow within the system.

14.3 SYSTEM CONSIDERATIONS

A diagram for a typical small computer is shown in Figure 14.6; the difference between a small calculator system and a small computer organization can be seen if this figure is compared with Figure 14.1. In a serial processor having the more complex structure shown in Figure 14.6, however, it is possible to use additional serial shift registers to control the memory cycle.

This technique has some disadvantages connected with speed, but it has the very great advantage that serial shift registers are indeed low-cost items. Using this technique, in fact, it would be possible to replace the entire memory block with shift registers. The more common implementation of Figure 14.6 is to use parallel organization in which each bus line shown

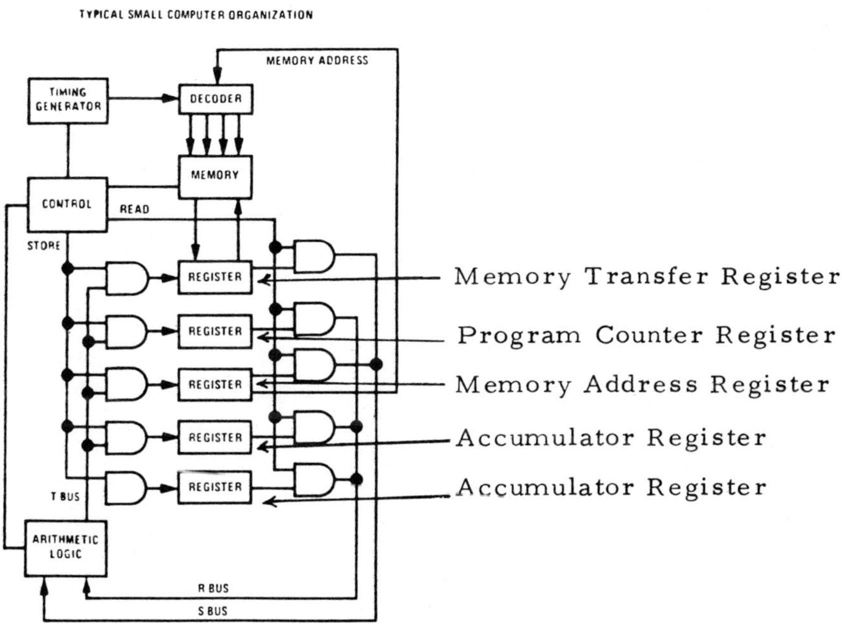

Figure 14.6. Typical small computer.

actually represents a number of leads equivalent to the number of bits in the data or instruction word. In both cases, however, direct microprogram control using read-only memories is practical.

In the small computer organization, the control unit receives an operation mode from memory, which enables it to place the proper signals on the store bus to correctly gate information into the active registers of the system from the *T* bus. Looking again at the ROM structure, one can see that the code from the main memory received by the control unit might be the input address to the read-only memory. Outputs from the read-only memory might represent the data flow control implemented along the store bus, with each output representing the function required at each storage-register input, depending upon the operation code.

The read-only memory can also be of great value as a memory-address decoder used to store information in the main memory. A specific advantage of this system organization is that it is not necessary to modify the control logic completely when it is desired to modify the function of a given system. It is only necessary to reprogram an additional set of read-only memories with the intent of changing the data flow and arithmetic logic operations as determined by the new set of operation codes. Therefore, it is easy to customize the application of the basic small computer organization by simply replacing read-only memory elements. The economic impact of this simple customization is very small, since the heart of the system remains unchanged.

15

MOS Main-Frame Memories

by Arturo Krueger

Motorola Semiconductor Products Division

By combining MOS and ECL (emitter-coupled logic) ICs, a versatile 8-*k* memory module was constructed for possible use as a building block for large systems. The complete 8-*k* memory, shown in Figure 15.1, is easy to use and has no complex timing requirements. When a binary coded address is applied to the inputs, the contents of the specified address appear at the output terminals within 120 ns and remain until a new address is presented. Writing in a desired location is accomplished by pulsing the write-enable line after the address and data inputs have been presented. The module can by cycled every 150 ns.

The memory module operates with +5- and —5.2-V power supplies. Inside the module, a variety of signal levels is encountered, reflecting the dual MOS–bipolar circuitry. The input–output levels, however, are compatible with standard emitter-coupled logic. The ECL circuits provide high-speed decoding and excellent drive capability. When a number of basic memory modules are to be connected into a large memory system, ECL provides the highest system speed. For computers operating with saturated logic, a simple translator can be provided at the memory system level. This will not be required in most cases, however, since the majority of present high-speed computers make use of some form of current–mode logic.

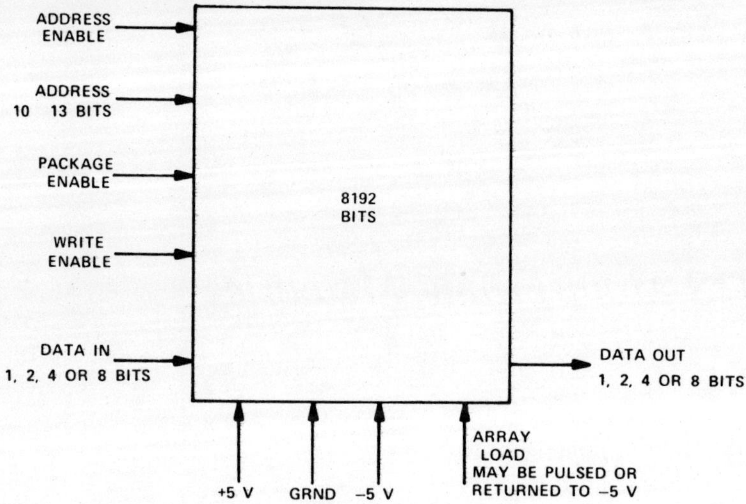

Figure 15.1. Block diagram for the 8192-bit module.

15.1 STORAGE ARRAY

The heart of any memory system is the storage array. The basic storage array used is a *p*-channel MOS array of 256-bit capacity. It is shown in block form in Figure 15.2. The chip is fairly large, measuring 138 × 141

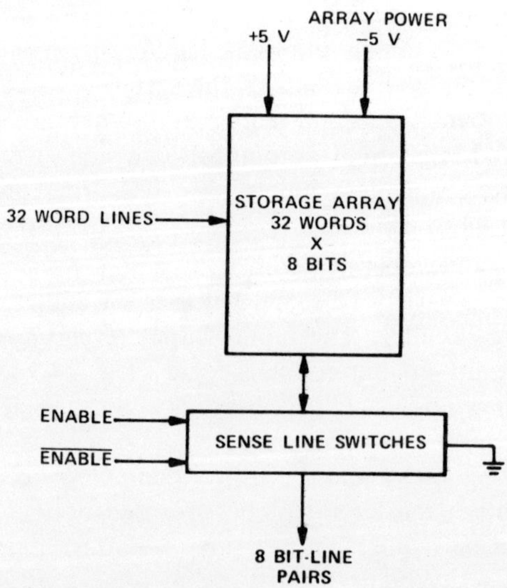

Figure 15.2. Block diagram for the MOS storage array.

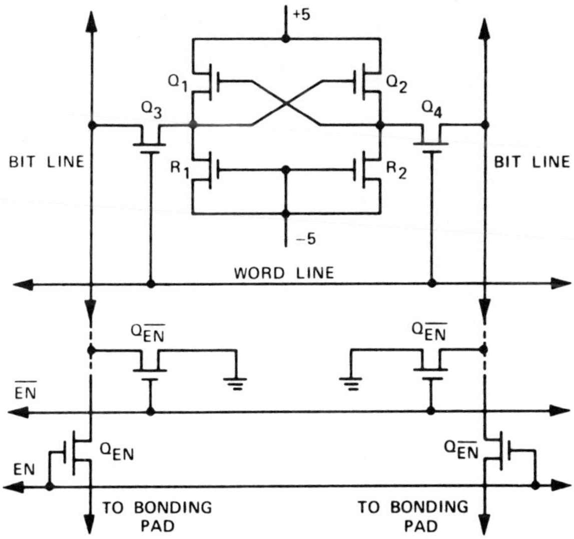

Figure 15.3. Circuit diagram for the storage cell.

mils. The diagram for a single storage cell is given in Figure 15.3. To minimize the on-chip interconnections and the number of devices per storage cell, a linear-select arrangement is used. That is, the array is organized in two-dimensional fashion as 32 words by 8 bits, as shown in Figure 15.4. The unusual technique of placing the sense line switches on the array chip is used in order to minimize the sense-loop time constant and improve the overall module decoding efficiency. The standby power dissipation of an individual storage chip is approximately 40 mW.

Eight of these 256-bit storage arrays are contained in an individual memory plane (a total of 2048 bits). The eight storage chips surround a

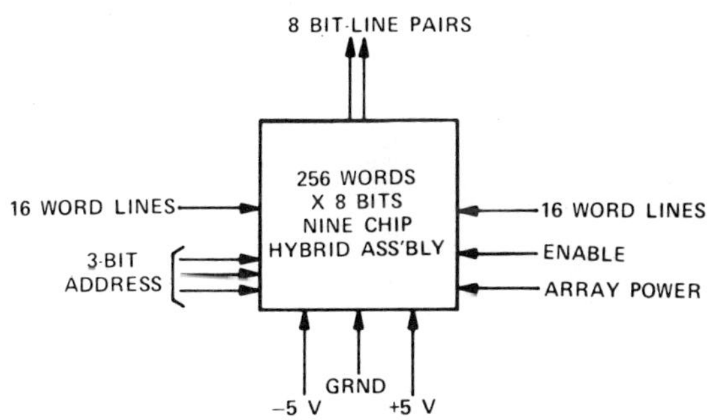

Figure 15.4. Memory package contains 256 words with 8 bits/word in a hybrid structure.

single bipolar circuit, the array-selection chip. The function of the array-selection chip is to "enable" the bit lines of one of the eight storage arrays. Voltage levels on the address inputs cause one of the array-selection circuit's eight output driver stages to provide the complementary signals necessary to drive the sense switches on the corresponding MOS storage array. Two additional binary coded inputs to the array-selection circuit

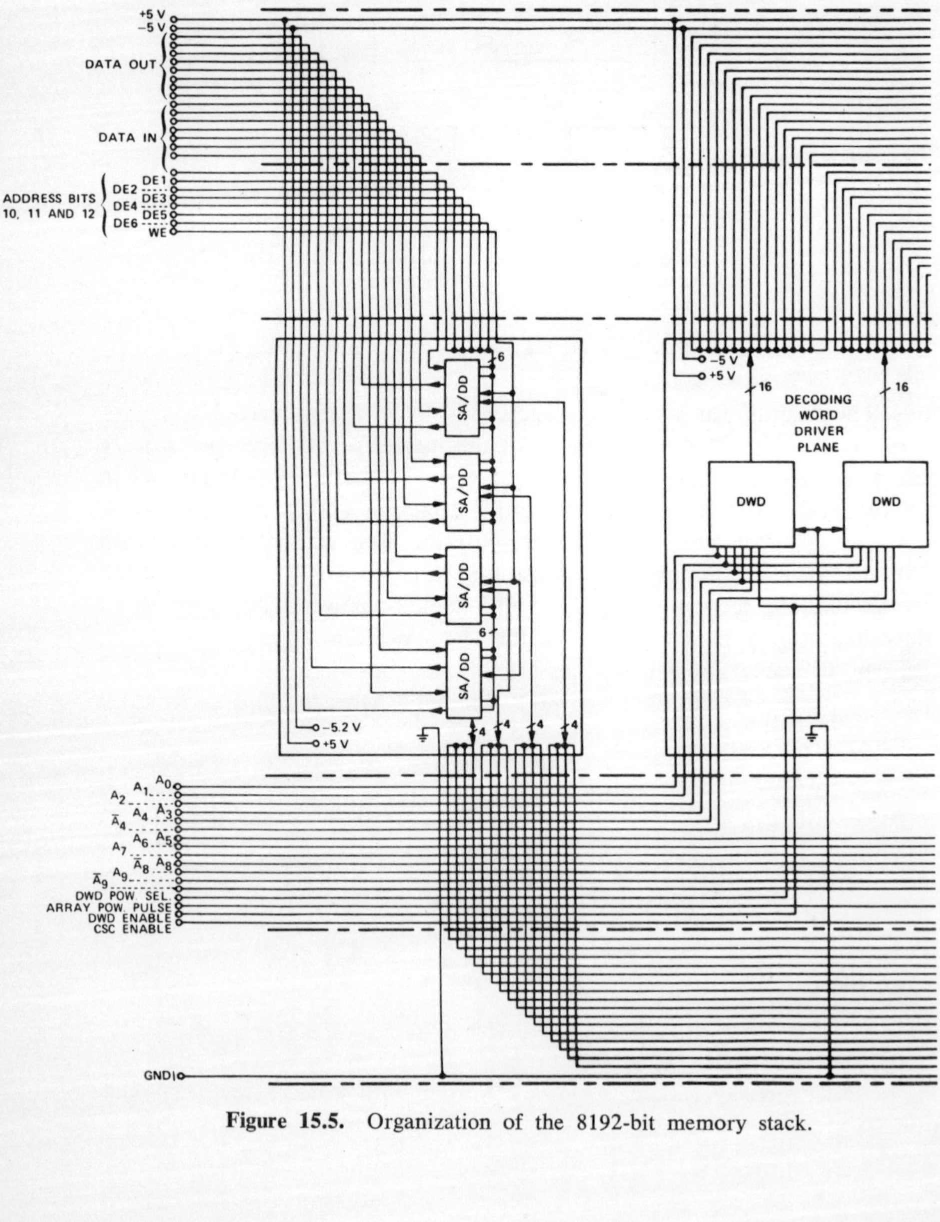

Figure 15.5. Organization of the 8192-bit memory stack.

allow the addressing of the array-selection circuits themselves, thus providing for selection of one of the four memory packages. The five inputs to the array-selection circuits, therefore, "enable" one of the four memory planes (a total of 8192 bits) and one of the eight storage arrays in the selected package. The entire 8196-bit organization is illustrated in Figure 15.5.

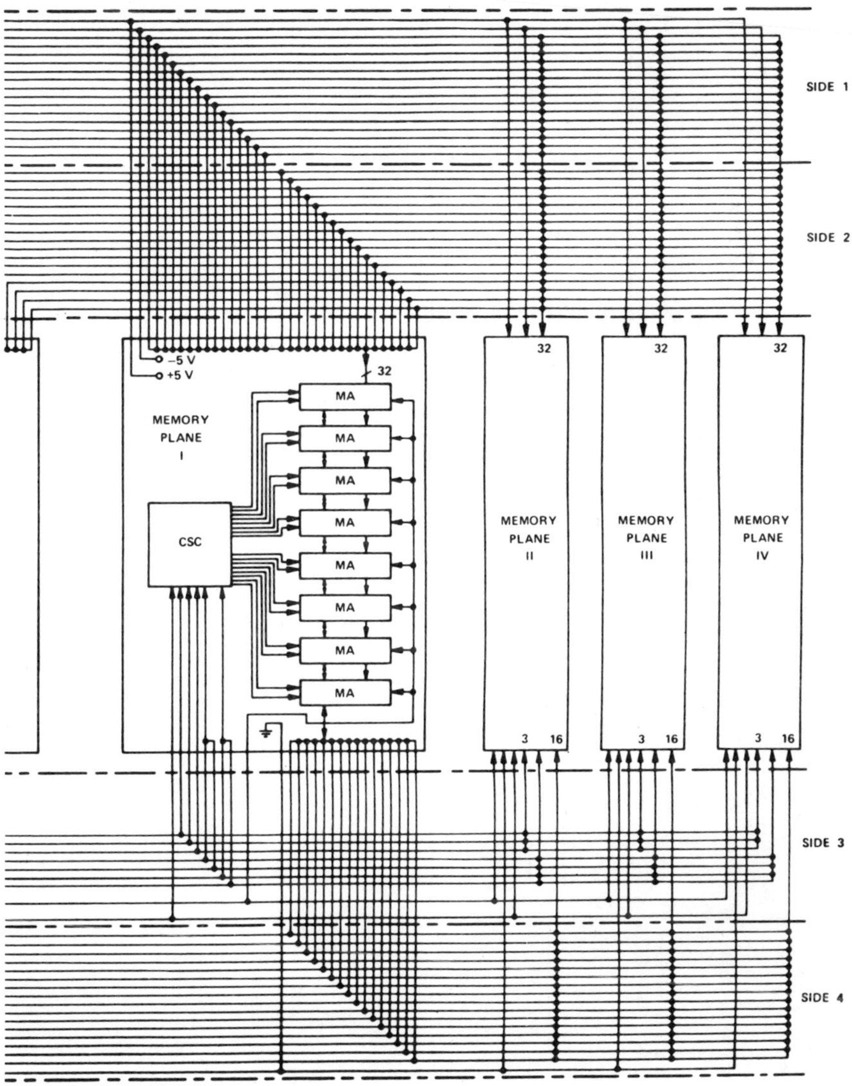

Figure 15.5. (*Continued*).

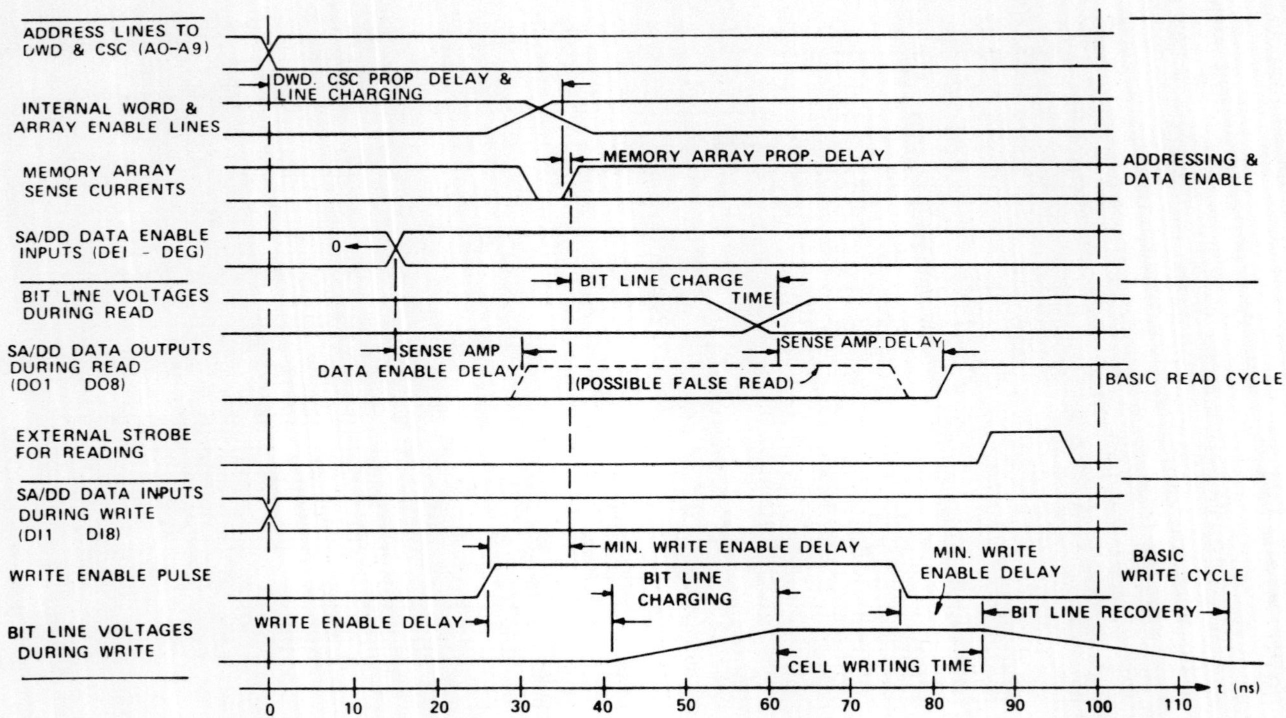

Figure 15.5. (*Continued*).

15.2 DECODING WORD DRIVER

The second basic unit in the module is the decoding word-driver package illustrated in Figure 15.6. Interconnection of the two bipolar, decoding word-driver chips (Figure 15.7) is made with a thick film metalization and conventional wire bonds. The package is similar to that of the memory-array package. The decoding word-driver chip logic is depicted in Figure 15.8.

Each of the two circuits decodes four address bits and drives one of 16 word lines. Two address-enable inputs are also provided. One is used as a master "enable" and the other acts as a 1-bit decoder to select one or the other of the two decoding word-driver chips sharing the package. Input signals to the decoding word-driver package thus select one of the 32 word lines common to all 32 storage-array chips in the module.

Selection of a single word in the memory modules is performed by: (1) selection of a memory plane and a single storage-array chip by the array-selection circuit in that plane and (2) selection of one of the 32 words on that storage-array chip by the decoding word-driver package. The chip-select circuit and the chip-select circuit logic are shown in Figures 15.9 and 15.10, respectively.

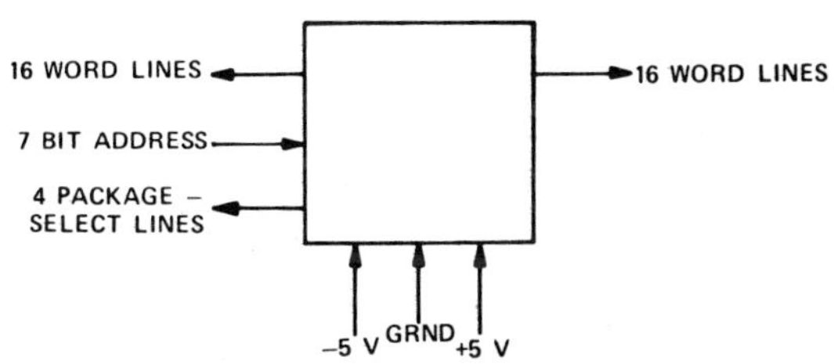

Figure 15.6. Decoding word-driver package.

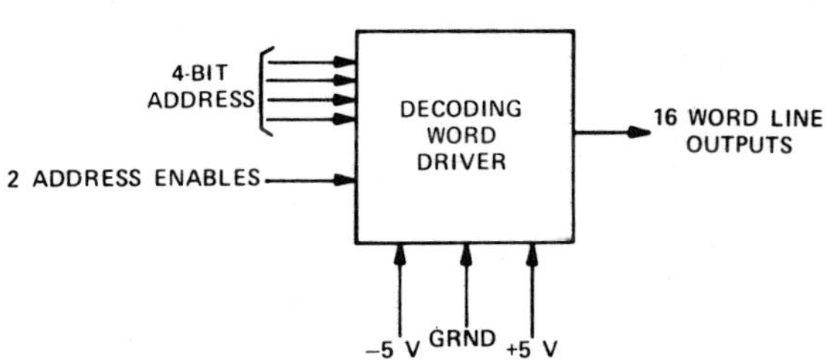

Figure 15.7. Decoding word-driver chip.

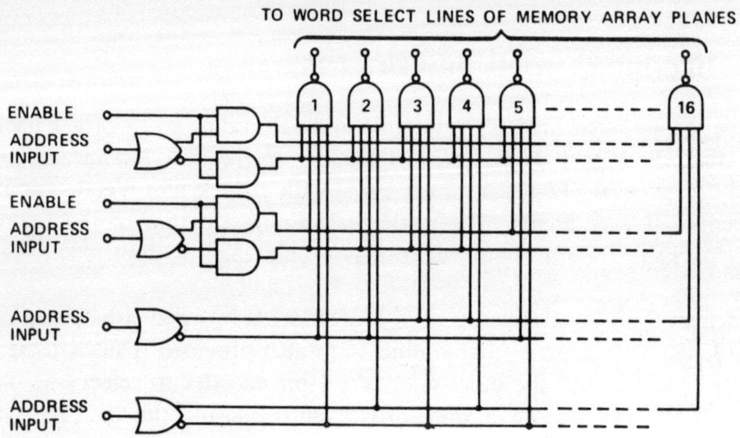

Figure 15.8. Decoding word-driver chip logic.

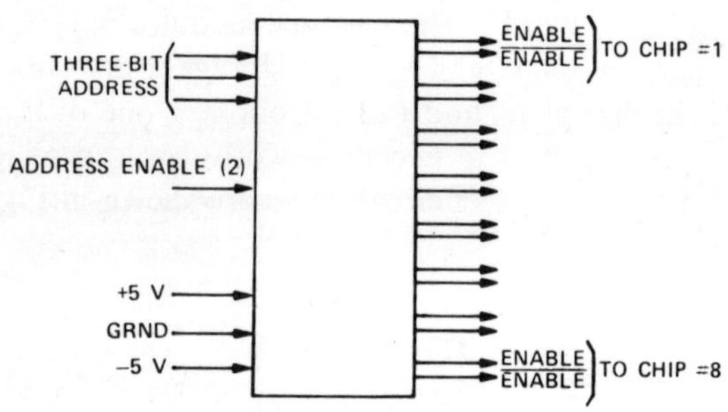

Figure 15.9. Chip-select circuit.

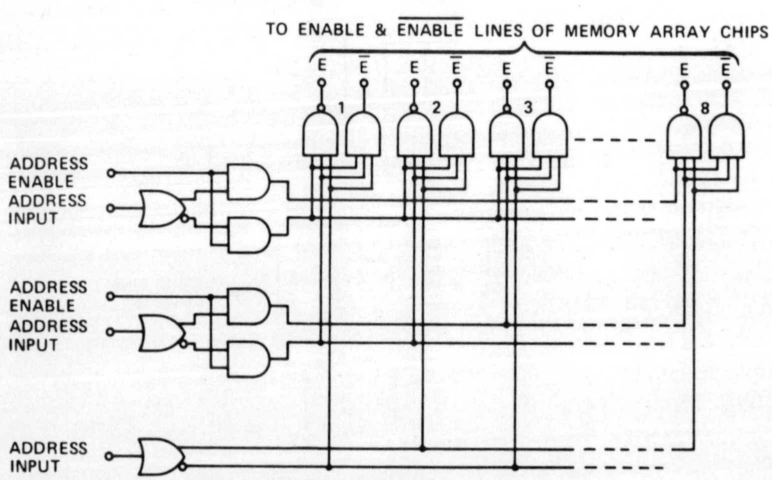

Figure 15.10. Chip-select circuit logic.

15.3 DATA INPUT–OUTPUT FUNCTION

The data input–output function of the 8 bit-line pairs in the module is provided by four identical, dual sense-amplifier/digit driver chips. The chips share one package and are interconnected, using the beam-lead laminate system.

Each chip serves two purposes. First, when properly "enabled" for writing, it transmits a write signal to appropriate bits of the selected word in the storage array. Second, when properly "enabled" for reading, it senses the MOS storage cell currents of the selected word and translates them to ECL-level signals at its data output terminals. Each chip therefore receives and sends MOS-level read and write signals to the storage arrays, accepts ECL-level data input and data "enable" signals, and generates ECL-level data output signals.

In the sense-amplifier–digit-driver circuit, in addition to the basic sense amplifier, digit (bit-line voltage) driver, voltage comparator, and gate blocks, there is a bit-line recovery circuit. The purpose of this circuit is to rapidly return all bit-line voltages to 0 after each write operation.

15.4 READING

Reading is performed by "enabling" either or both halves of the sense-amplifier–digit-driver chip with the data enable signals. If the write enable is held at a logical "1," the bit-line currents flow into the sense-amplifier inputs. The sensed information is made available at the data-out terminals. Because the I/O signals are at ECL levels, uncommitted emitter outputs are used so that wired ORing of the positive-going output signals is possible.

The sense–digit package is shown in Figure 15.11.

Since the data-out signals from all four sense-amplifier/digit-driver chips can be ORed, various connections of the data-enable and data-in signals are possible. If the data enables are connected for maximum decoding, a 1-out-of-8 selection can be accomplished. With all eight data-in inputs and data-out outputs strapped together, the module organization becomes 8192 words of one bit. Similarly, if all data enables are tied together, each data in and data out is used as a separate information channel and the resultant module organization is 1024 words of eight bits. Other connections result in 4096 words of two bits and 2048 words of four bits. These connections are made external to the module.

The package-to-package connections are accomplished by means of parallel buses or rails, formed when the six packages are stacked. Although

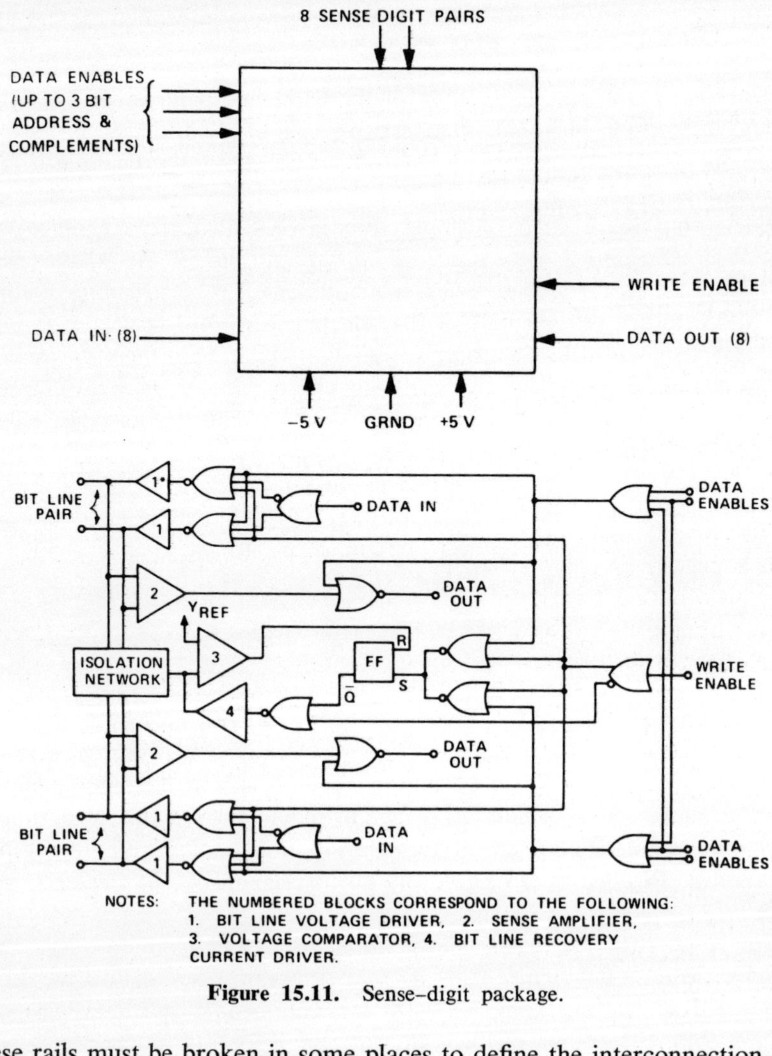

Figure 15.11. Sense–digit package.

these rails must be broken in some places to define the interconnection, no jumper wires need be added.

15.5 TIMING

For purposes of discussion, the 8-*k* module timing can be divided into three sections: the basic addressing and enabling common to both read and write cycles, the basic read cycle, and the basic write cycle.

If desired, the sense-amplifier–digit-driver data-enable inputs (DE_1 through DE_6) can arrive at the module terminals some 20 ns later than the address signal (A_0 through A_9). Therefore, additional levels of decoding logic may be added in series with these inputs without affecting the cycle time.

During the read cycle, the sensed information must be strobed out of the module toward the end of the cycle, because the information on the sense-amplifier–digit-data outputs prior to that time is either the stored information in the previously addressed location or undefined. Bit-line recovery time after writing overlaps into the next cycle.

Waveform photographs of a partially populated module of an 8192-bit system show that the address may be changed every 150 ns for reading or writing in any sequence. Read-access time when preceded by a read cycle is about 90 ns. Read-access time when preceded by a write cycle is also about 90 ns.

The capacitances associated with the word selection and bit lines are high (approximately 250 pF on word lines, 70 pF on bit lines), since these lines are paralleled with all 32 storage-array chips. Charging times, therefore, are a significant part of the cycle time.

The bipolar driving and sensing circuits were designed to minimize the delays caused by charging and discharging these distributed capacitances. Specifically, the output stages of the array-selection and decoding word-driver circuits provide sufficient transient current to rapidly charge and discharge the distributed capacitance. Very little dc drive (1 to 2 mA) is required.

15.6 SENSING SMALL CURRENTS

In order to sense small currents from the MOS storage cells, the termination resistance for the bit lines at the inputs to the sense amplifier–digit drivers must be very low. The resistances at the terminating nodes, however, could not be made small enough to avoid a relatively long time constant because of the associated 70-pF distributed capacitance. The sense-amplifier–digit-driver design, therefore, features an input stage that will not only detect small differential bit currents (20 μA) with common-mode rejection of +500 to —250 mV, but will, after the detection voltage threshold has been exceeded, clamp the differential voltage of the bit lines at a low level (about 30 mV) for a wide range of cell currents. Consequently, very little charging and discharging of bit-line capacitance occurs during a read cycle.

15.7 WRITE-RECOVERY CIRCUIT

Another feature of the sense-amplifier–digit-driver circuit is the use of a write-recovery circuit that is common to both bit-line pairs serviced by the circuit.

Although differential bit-line voltage is limited during a read cycle, it is purposely made large (about 4 V) during the write cycle to guarantee writing into the MOS storage cell. As previously mentioned, the bit-line pairs normally have a large time constant associated with them, and to allow the bit line that is driven high during writing to recover unaided from the 4-volt point-to-ground potential would require a prohibitively large amount of time. The problem is minimized, however, by the action of the recovery circuit, which quickly pulls the bit line back to ground potential following a write operation. Gating circuitry is also included on the sense amplifier turned off during the write operation in order to minimize the loading of the digit-driver circuit.

16

Bipolar Random-Access Memories

by Melvin G. Snyder

Raytheon Company

The first monolithic semiconductor bipolar trip to be widely used as the basic building element for packaging 256-bit, 512-bit, and 1024-bit random-access-memory arrays was the 64-bit chip.

16.1 ORGANIZATION OF MEMORY

This high-speed TTL random-access-memory element is organized in a linear-select 16-word $\times$ 4-bit array. On-chip address decoding along with chip-enable, write-enable, and uncommitted-collector outputs provide for simplified connection of these memory elements into larger arrays. The entire chip, including address decoding and write and sense circuitry, typically dissipates 350 and 420 mW, maximum. A word is accessible in less than 45 ns, the write-recovery time is less than 35 ns, and the minimum write pulsewidth required is less than 30 ns. The entire chip measures approximately 90 $\times$ 105 mils, depending on the manufacturer, and uses single-level metal interconnection.

Each one-bit memory element is a latch flip-flop. This simple form of flip-flop used for bit storage is composed of a pair of two-emitter transistors and three resistors. One emitter on each transistor is used for word selection. One emitter remains at a reference voltage of 1.7 V, and the remaining emitter is used for writing information into the cell and reading information from the cell (see Figure 16.1).

The monolithic memory cell also contains the write enable that controls data to be written into the flip-flop, allowing only selected data into the cell when its input is "low" (greater than 1.7 V). No data can be written into the flip-flop when the input to the write enable is high (less than 1 V). Conversely, no data can be read out of the flip-flop when the write-enable input is "low." The write amplifier will write selected data into the flip-flop only when the write-enable input and the chip enable are "high" and the input is "low." The sense amplifier will read selected data out of the flip-flop only when the chip-enable is "low" and the write-enable input is "high."

To analyze the read operation in more detail (see Figure 16.2), assume that a logical "1" is stored in the flip-flop. This means that transistor Q_1 is "on" and conducting current. As long as the word-selection line is "low" (less than 1 V), the current path will be through this line. The word-selection line "low" is actually 0.5 V. However, if the word-selection line is "high" (greater than 1.7 V) and the write-enable line is "high," the current will switch to the second emitter and flow down into the appropriate data sense amplifier, activating a logical "1" at the data output of this amplifier.

Now, let us examine the write operation in more detail by writing a logical "0" into the flip-flop. First, chip enable and write enable must be "low" (greater than 1.7 V), and the cell is selected through the appropriate binary code at the address inputs. Then, the input to the selected write amplifier is reduced to a "low" (less than 1 V), turning "off" the output of the amplifier. This will, in turn, raise the voltage on the write/read line to a "high" greater than 1.7 V), causing transitors Q_1 and Q_2 to turn "off" and "on,"

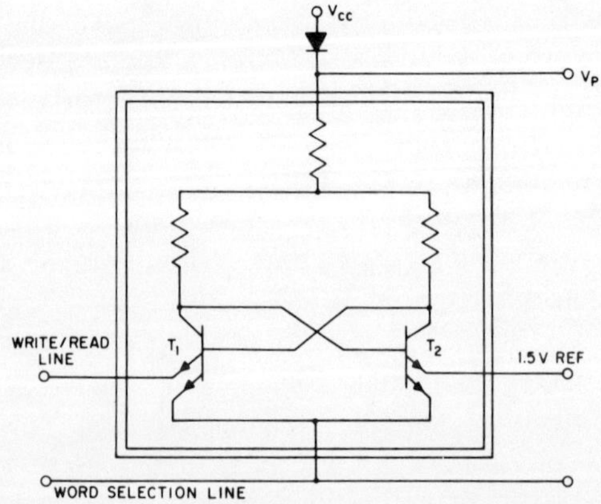

Figure 16.1. Diagram of bipolar memory cell.

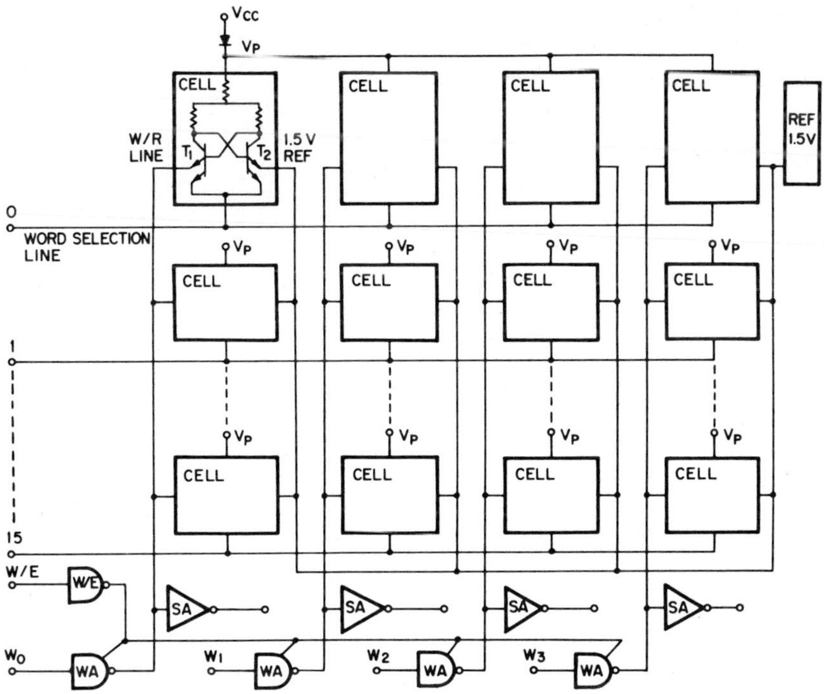

Figure 16.2. Sixteen-word, 4-bit bipolar memory.

respectively. Now the flip-flop is said to have a logical "0" stored. In order to write a logical "1" into the flip-flop, the chip enable and write enable remain the same but in this case, the input to the selected write amplifier is increased to a "high." This will in turn lower the voltage on the write/read line to a "low," causing transistors Q_1 and Q_2 to turn "on" and "off," respectively.

A reduction in the number of pins required for a memory element can be accomplished by the addition of an address decoder (see Figure 16.3). The addition of this decoder reduces the pin count of this 16-word, 4-bit memory element from 28 pins to 16 pins, including chip enable. This decoder has five inputs. Four of the inputs, A_0, A_1, A_2, and A_3, act as addresses to produce an output at a corresponding output terminal; the fifth input, chip enable, is used primarily in larger memory arrays to select a 64-bit memory-element chip. The A_0 input terminal of the decoder is considered the least significant, so that an input configuration such as "1000" on inputs A_0, A_1, A_2, and A_3, respectively, will produce a logical "1" on output 1.

To analyze the address decoding operation in more detail, place a logical "0" (less than 1 V) at the chip-enable input and A_1, A_2, A_3 inputs while

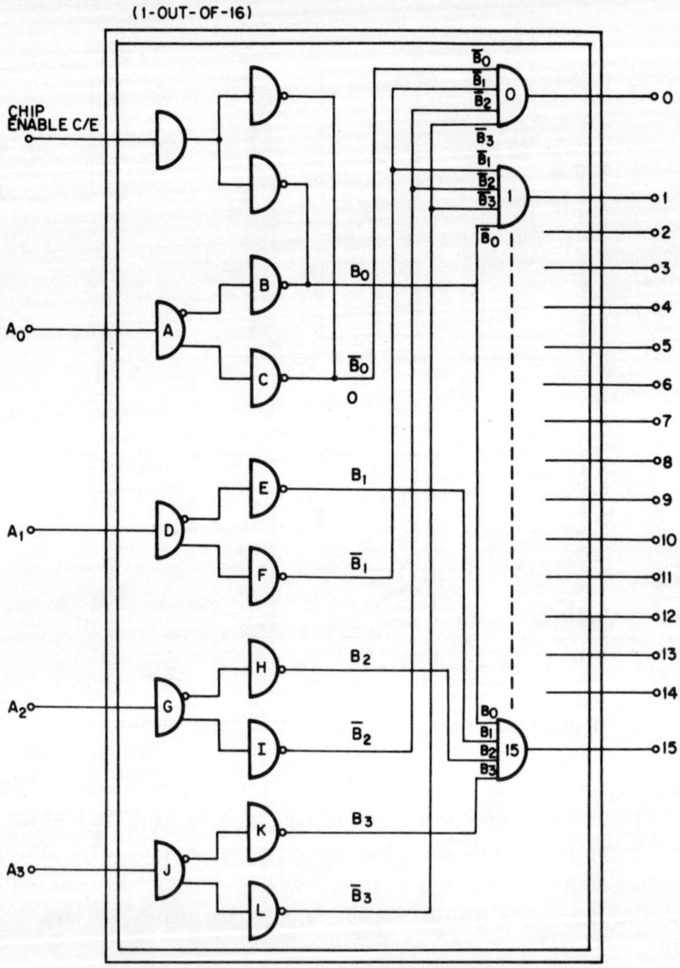

Figure 16.3. Diagram of 1-out-of-16 decoder.

placing a logical "1" (greater than 1.7 V) at the A_0 input, 1000. This will cause gate B to turn "off" and gate C to turn "on," resulting in logical "1"s at all the inputs of gate 1. Gate 1 will now be turned "off," resulting in a logical "1" at output 1, while all the other gates, 0 and 2 through 15, have at least one logical "0" at the input, leaving their outputs at logical "0" (see Figure 16.4).

Looking further, the truth table shows the A_3 input terminal of this decoder to be the most significant bits, so that an input configuration such as "1000" on inputs A_0, A_1, A_2, and A_3, respectively, would produce a logical "1" on output 1; the A_3 input terminal of this decoder is seen to be the most

			Input											Output					
A_0	A_1	A_2	A_3	0	1	2	3	4	5	6	7	8	9	10	11	12	13	14	15
0	0	0	0	1	0	0	0	0	0	0	0	0	0	0	0	0	0	0	0
1	0	0	0	0	1	0	0	0	0	0	0	0	0	0	0	0	0	0	0
0	1	0	0	0	0	1	0	0	0	0	0	0	0	0	0	0	0	0	0
1	1	0	0	0	0	0	1	0	0	0	0	0	0	0	0	0	0	0	0
0	0	1	0	0	0	0	0	1	0	0	0	0	0	0	0	0	0	0	0
1	0	1	0	0	0	0	0	0	1	0	0	0	0	0	0	0	0	0	0
0	1	1	0	0	0	0	0	0	0	1	0	0	0	0	0	0	0	0	0
1	1	1	0	0	0	0	0	0	0	0	1	0	0	0	0	0	0	0	0
0	0	0	1	0	0	0	0	0	0	0	0	1	0	0	0	0	0	0	0
1	0	0	1	0	0	0	0	0	0	0	0	0	1	0	0	0	0	0	0
0	1	0	1	0	0	0	0	0	0	0	0	0	0	1	0	0	0	0	0
1	1	0	1	0	0	0	0	0	0	0	0	0	0	0	1	0	0	0	0
0	0	1	1	0	0	0	0	0	0	0	0	0	0	0	0	1	0	0	0
1	0	1	1	0	0	0	0	0	0	0	0	0	0	0	0	0	1	0	0
0	1	1	1	0	0	0	0	0	0	0	0	0	0	0	0	0	0	1	0
1	1	1	1	0	0	0	0	0	0	0	0	0	0	0	0	0	0	0	1

Figure 16.4 Truth table for 1-out-of-16 address decoder.

significant bit, so than an input configuration such as "000" would produce a logical "1" at output 8.

Now that the various sections of the 64-bit memory array have been discussed, the entire element can be connected together (see Figure 16.5). To read the sixteenth word out of this 16-word, 4-bit memory element, the chip-enable and the write-enable inputs must be placed at logical "0" and all inputs A_0, A_1, A_2, and A_3 must be at a logical "1." This code will cause the correct word selection line (15) to go "high," resulting in data from all four bit locations of the sixteenth word to be read out simultaneously at D_0, D_1, D_2, and D_3 outputs.

To write data into the ninth 4-bit word of this memory element, the write-enable input must be placed at logical "0," the chip-enable and write-amplifier inputs must be at logical "0," and the inputs A_0, A_1, A_2, and A_3, must be placed in logical 1001, respectively. This configuration will cause the correct word-selection line (8) to go "high," allowing data at the write inputs (W_0, W_1, W_2, and W_3 to be written into the four bit locations of this word.

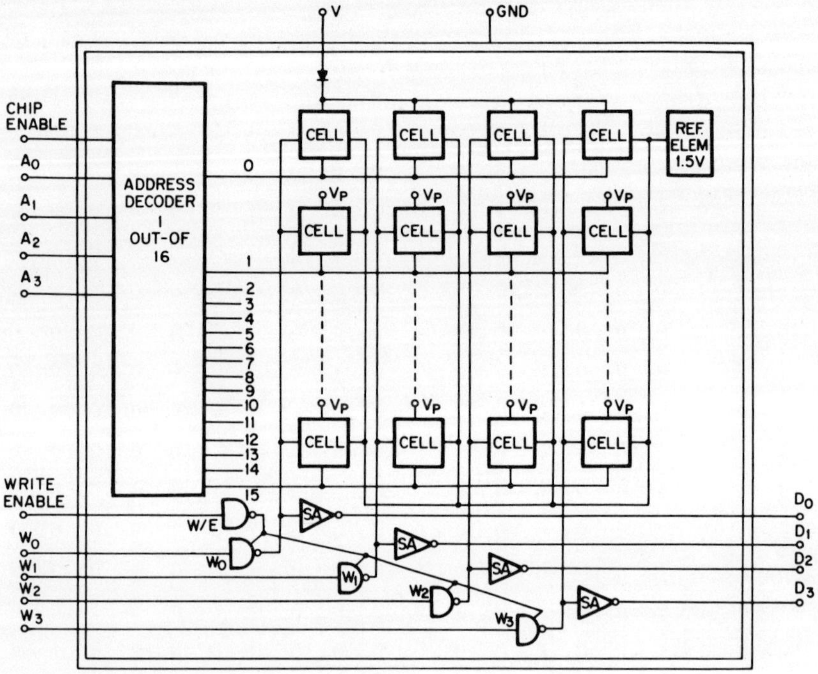

Figure 16.5. Diagram of the 64-bit memory.

16.2 LSI MEMORY SUBSYSTEMS

By combining monolithic semiconductor memory technology with multichip beam-lead packaging techniques, low-cost, small-size, highly reliable LSI memory subsystems can be built. Individual 64-bit memory elements can be bonded on a substrate to gold-plated interconnected metalization paths within a 1×1 in. package. This modular approach allows storage capacity to be increased in an orderly evolutionary manner without any degradation in speed.

This technique can be used to build a 1024-bit RAM subsystem. This memory subsystem is constructed by bonding sixteen 64-bit (16 word, 4 bit) beam-lead memory-element chips to an interconnected metalized substrate.

LSI memory subsystems have been built organized as linear-select 256-word $\times$ 4-bit and 128-word $\times$ 8-bit arrays. Numerous other configurations can be built—any memory N times 16 words by M times 4 bits; up to thirty-six 64-bit memory elements can be ORed together without additional gates. The high-speed 256-word, 4-bit memory subsystem is constructed by bonding sixteen 16-word, 4-bit memory elements to a metalized substrate that connect all the write-amplifier inputs together, all the data outputs together, all the address lines together, and all power supply lines and ground lines together; it also connects the appropriate external pins (see Figure 16.6). A 1-out-of-16 decoder chip may be added to allow for binary inputs.

To make a high-speed 128-word, 8-bit memory subsystem, sixteen 16-word, 4-bit elements are bonded to a metalized substrate. The substrate interconnection pattern connects the address lines together and the write-enable lines together. The pattern also connects together all other memory element data outputs; it connects all power supply and ground lines together, and connects the appropriate external pins (see Figure 16.7). A 1-out-of-8

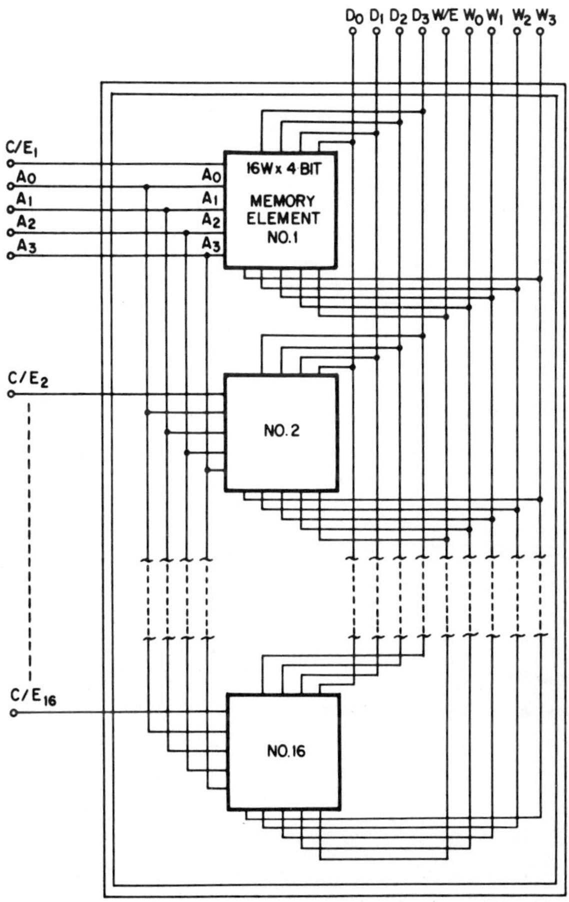

Figure 16.6. Diagram of 256-word, 4-bit memory subsystem.

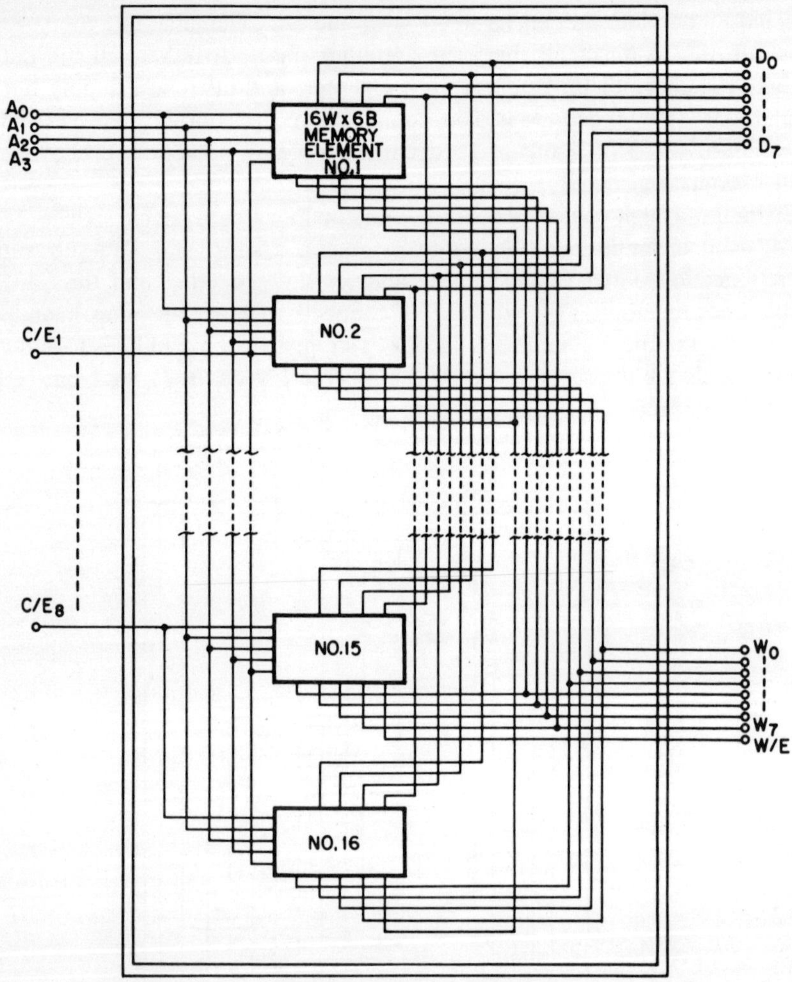

Figure 16.7. Diagram of 128-word, 8-bit memory subsystem.

decoder chip may be added to allow for binary inputs. The power dissipation of an element is approximately 250 mW, including address decoding and write and sense circuitry, whereas the entire power dissipation of either memory subsystem is about 4 W. A word is accessible in less than 45 ns, the write-recovery time is less than 35 ns, and the minimum write pulse-width required is less than 30 ns. A 1024-bit memory subsystem requires 31 external pins to operate, in contrast to a 64-bit (16-word, 4-bit) memory element where 16 external pins are needed. A 2^4 increase in memory bit size does not quite double the amount of pins needed to operate the 1024-bit memory subsystem (see Figure 16.8).

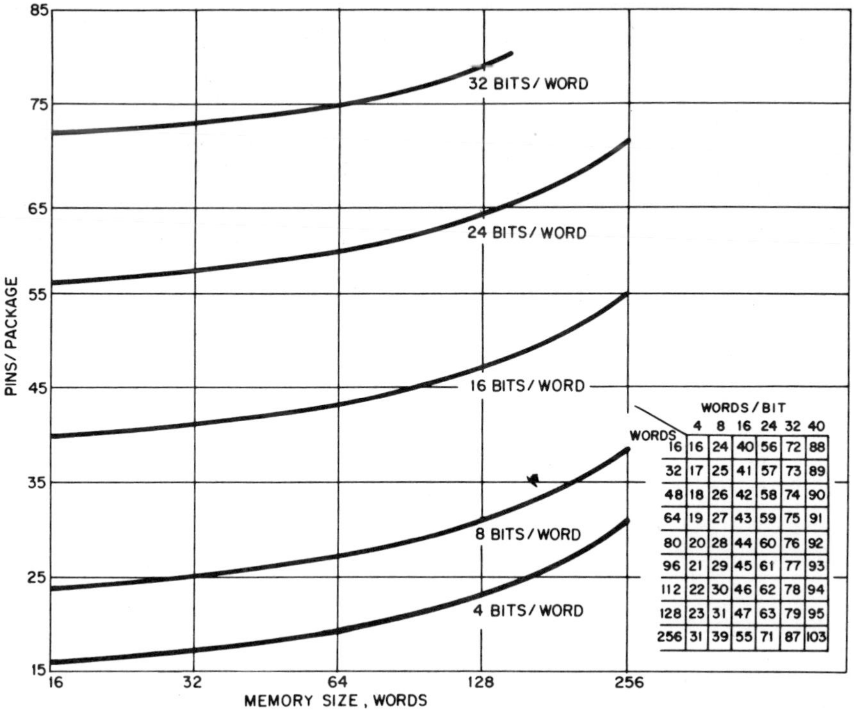

WORDS \ WORDS/BIT	4	8	16	24	32	40
16	16	24	40	56	72	88
32	17	25	41	57	73	89
48	18	26	42	58	74	90
64	19	27	43	59	75	91
80	20	28	44	60	76	92
96	21	29	45	61	77	93
112	22	30	46	62	78	94
128	23	31	47	63	79	95
256	31	39	55	71	87	103

Figure 16.8. Memory size versus pins per package.

Figure 16.8 also shows that the rate at which the number of bits per word increases has a greater effect on the rate with which the number of external pins increases than it does on the rate at which the number of words increases. Therefore, to minimize the number of external pins required for a given configuration, the bits per word should be as low as possible. Modular organization of these memory subsystems facilitates the construction of larger memories. The major limitation in the packing density of these larger memories is determined by the power dissipation, or more specifically, by how the memory will be cooled.

16.3 APPLICATIONS

In normal operation of computers and similar equipment, the location-addressable form of random-access memories appears to have the widest range of applications. The most common of these are scratch-pad memories, in which information is temporarily stored. A typical example is the use of a scratch-pad memory to store the intermediate results of the arithmetic or logical operations, instead of storing these intermediate results in the main memory and later fetching them as required, which causes long delays.

Another example is the use of a location-addressable memory to store the state of the machine quickly in preparation to service and interrupt. This is particularly useful in a multiprogram environment.

Other applications include book marking, main-memory program relocation control, look-ahead programming, memory protection, and input-output data-transfer buffering.

17

Designing with Bipolar Memories

by Anh Nguyen-huu

Transitron Electronic Limited

The 64-bit bipolar memory IC is a word-addressed element, arranged in a 16 $\times$ 4 matrix with internal (on-chip) decoding facility. The storage unit is arranged as a 16-word, 4-bit block. The memory bits are formed by a pair of two-emitter transistors, cross-coupled to provide the basic set–reset flip-flop configuration. This device also includes four sense amplifiers, each of which is arranged to indicate the state of the selected bit. The sense amplifiers enable the four bits of a word to be read simultaneously. Four write amplifiers are also provided to allow any desired information pattern to be written into the four corresponding bits of the selected word. In addition to this, an address-select-enable amplifier is incorporated to provide inhibition of the address selection when not required, and a write-enable amplifier is included to provide the write command signal.

The number of pins required for a 64-bit memory, arranged as a 16-word, 4-bit element without on-chip decoding networks would be at least 22. However, when decoding is added to the device, the total number of pins required is 16. The decoder requires five inputs, four of which, A_1 through A_4, are address inputs, whereas the fifth one is used as an address-enable input. This input is particularly useful in large memory arrays. The sense amplifiers have open collector outputs, so that many sense amplifiers of a large memory array can be tied together, forming a wired-OR function. Commercially available 64-bit memories are designed to operate from a power supply of 5 V, and their inputs and outputs are fully compatible with both DTL and TTL ICs.

The threshold levels of this device are within the normal TTL threshold range, and the input-decoding networks contribute another advantage to those already discussed, since they help to reduce the input load current to 1.4 mA maximum.

17.1 BASIC ELEMENT

The basic storage section for the entire memory elements is a block of 64 set–reset flip-flops, arranged, as previously mentioned, in a 16 $\times$ 4 matrix. Each flip-flop is composed of a pair of two-emitter transistors and three resistors. One emitter from each transistor of a flip-flop bit is tied together and is used for word selection. As shown in Figure 17.1, one emit-

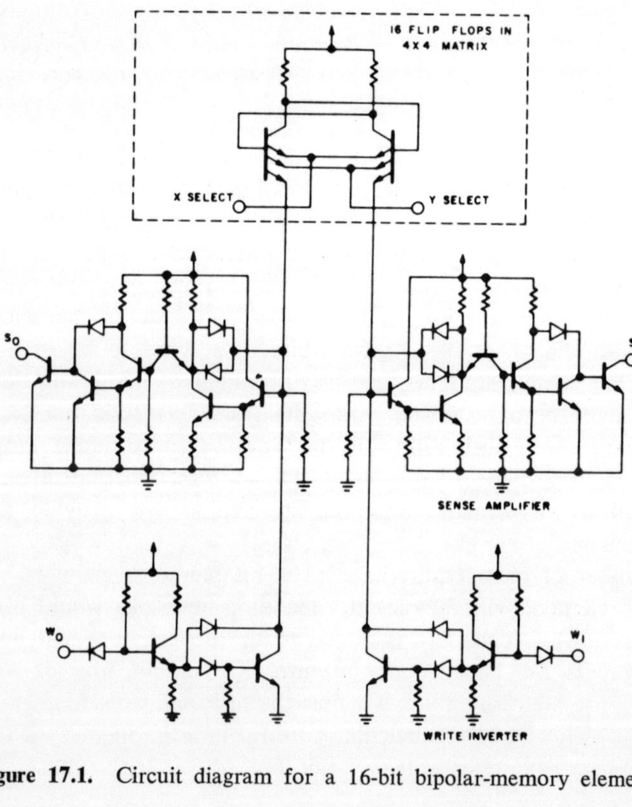

Figure 17.1. Circuit diagram for a 16-bit bipolar-memory element.

ter is used as a sense–write line and the other is connected to a reference voltage level which is the same for all 64 cells. The resistors of the basic flip-flops are arranged in a special manner to provide better threshold sensitivity, and they are of high values to reduce total device power dissipation. The selection of the 4-bit words is achieved by means of the 16 decoding four-input transistors, which are also parts of the input network.

17.2 ADDRESS DECODING

Four address buffers with complement output are used to provide the selection signals that are fed to the input of the 16 four-input transistors. Because only one of the 16 word lines is to be selected at any time, a special circuit arrangement is used for these buffers so that they provide the necessary overlap, thus eliminating the possible multiple selection of the word lines. As shown in Figures 17.2 and 17.3, address input A_1 is specially ORed with the invert signal of the address selection enable input. If the address-enable input (AE) is brought to a logical "1" level, both the A_1 and the A_1' outputs of the address buffer a_1 are saturated. Since the 16 possible logic combinations of the four address inputs A_1 through A_4 contain either the signal A_1 or A_1' when both are "0," no word selection is possible. Therefore, an address selection is active only when the address-enable input is connected to a logical "0" level.

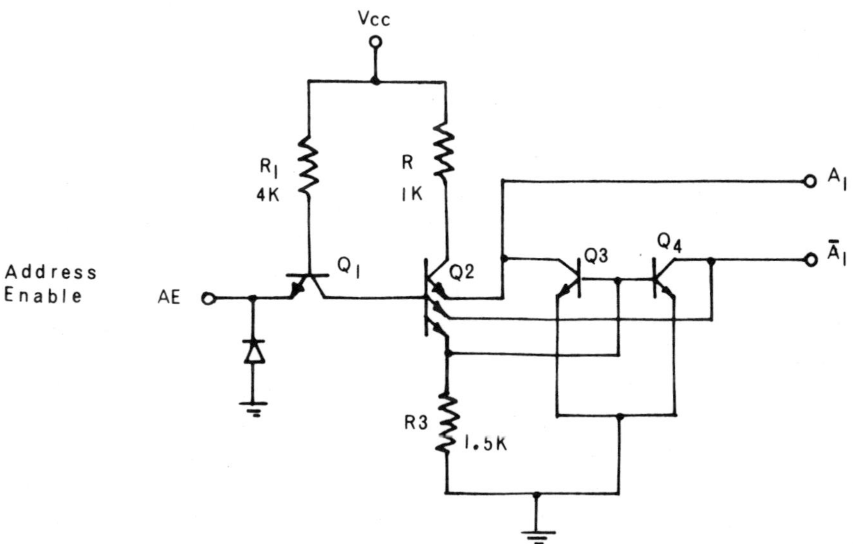

Figure 17.2. Circuit diagram for the address-enable network.

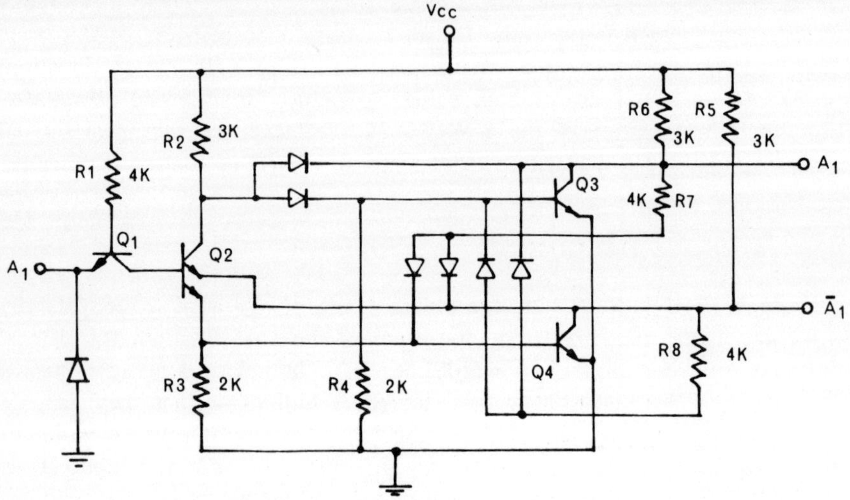

Figure 17.3. Circuit diagram for the address input.

17.3 SENSE AND WRITE AMPLIFIERS

The output of a sense amplifier changes its state according to the content of the bit location under observation. The inputs of the sense amplifiers are kept at a constant voltage level of V_{ref}, which is maintained by a special network. These sense amplifiers work as current detectors with their inputs set, and the sense emitter of the selected bit location remains at a constant level of approximately 1.5 V. If the current level of the sense line increases, the output transistor of this sense amplifier will be off. However, no change in potential is presented to the set–reset flip-flop, and therefore, the information previously stored in this bit location will not be destroyed during the sensing process (see Figure 17.4).

In the writing arrangement, data inputs are ANDed with the complement of the write-enable input. A writing operation can only be achieved when the write-enable input is connected to a low potential. Assuming that the flip-flop shown in Figure 17.1 is one of the 4-bit locations selected, and that the transistor Q_1 is on (information content is "0"), the base of this transistor is held at a potential of $V_{\text{ref}} + V_{\text{CE}}$ (Q_2) while its emitter, which is also the sense line, is held at 1.5 V. When the address-enable input is connected to a low potential, the word line is selected, and when the data input is connected to a potential of less than 0.8 V, the output of the write amplifier is off. The sense line potential is therefore raised to a potential of $V_{\text{ref}} + 1V_{BE}$ and the transistor Q_1 will turn off, forcing the other transistor Q_2 to be on.

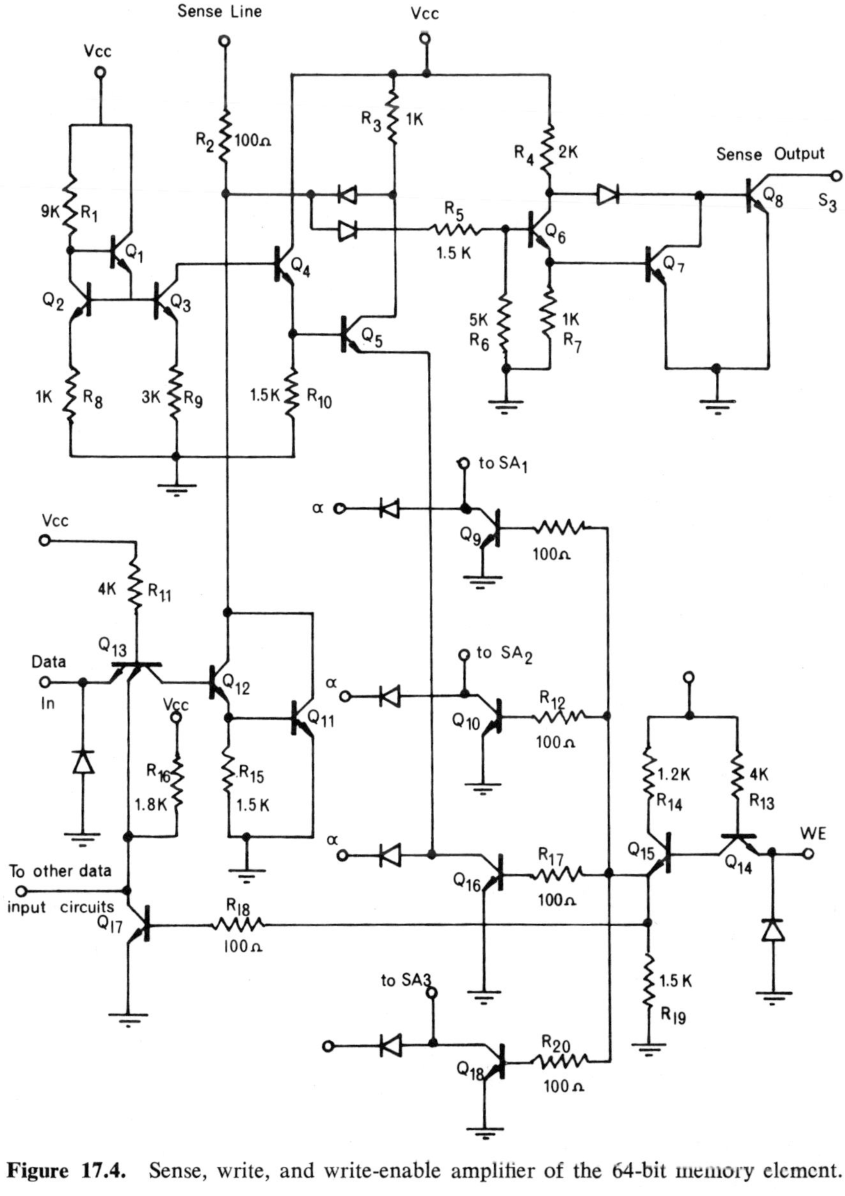

Figure 17.4. Sense, write, and write-enable amplifier of the 64-bit memory element.

If the data input is now connected to a potential of more than 2.2 V, the output of the corresponding write amplifier is on, pulling the sense line down $1V_{BE}$ below the reference level. Since Q_2 is on, the base of Q_1 is held at $V_{ref} + V_{sat}(Q2)$, with its emitter now lowered to $V_{ref} - 1V_{BE}$. This transistor will again turn on, thus causing Q_2 to stop conducting. Only one write amplifier is used, either for the writing of a logical "0" or a logical "1" to the words per bit selected.

17.4 CIRCUIT OPERATION

The complete circuit configuration is shown in Figure 17.5. The selection of a word line, say the eighth line, is achieved by applying signals (0001) to the four address inputs, A_1 through A_4 and by connecting the address-select enable to a logical "0" level and the write-enable input to a

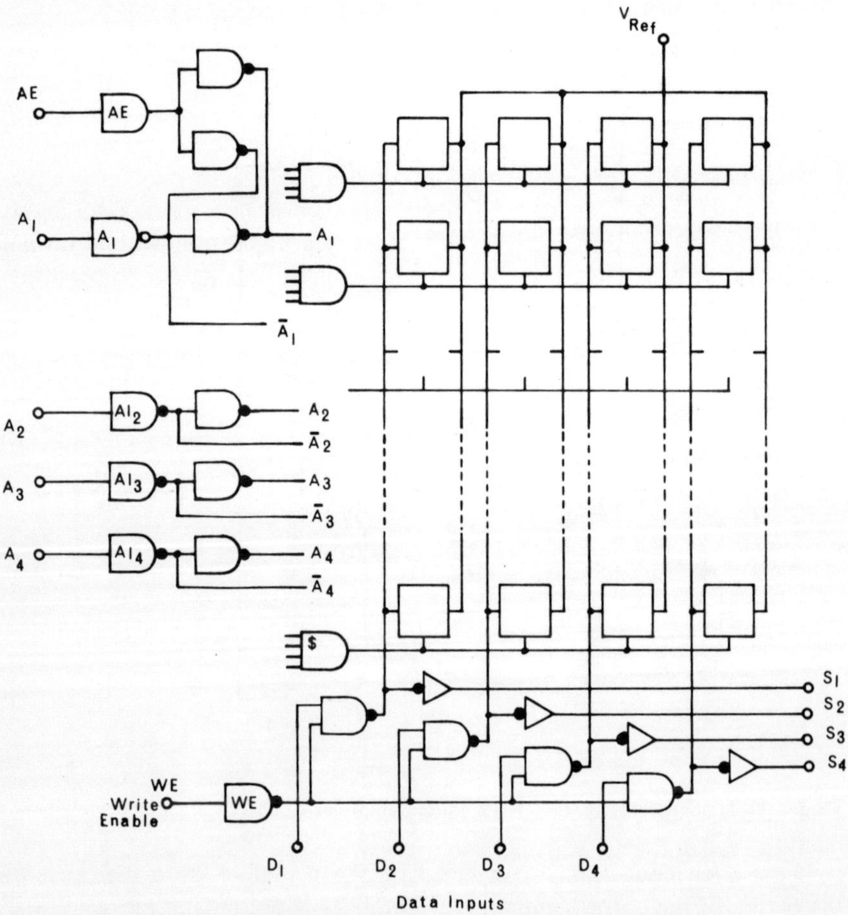

Figure 17.5. Functional diagram for the 64-bit memory element.

logical "1" level. If any of the four bits of the selected word line has a content of "1," the corresponding sense-amplifier output will turn on. If any of those four bits has a content of "0," the corresponding sense-amplifier output will remain off, which is also the state of the sense-amplifier outputs when no selection is made.

The writing of a signal pattern into the four bit locations of a word line, say the tenth line, is achieved by applying signals (0101) to the address inputs A_1 through A_4, by connecting the address-enable input to a low potential to enable the selection, applying the signal pattern to be written, to the four write-amplifier inputs, and lowering the write-enable input to a low potential. A logical "0" or a logical "1" level will be written into the selected bit or bits, depending on whether the signal applied to the write imputs is a logical "0" or a logical "1." The write amplifier only comes into action when a logical "1" level is to be written.

17.5 MULTICHIP ARRAYS

Two 64-bit memory elements can be grouped together to form a 16-word × 8-bit or a 32 word × 4-bit memory array. The appropriate inputs must be ORed together. These forms of array are quite simple and do not really need internal decoding; 24 and 17 contacts to outside, respectively, are required by the substrates (see Figures 17.6 and 17.7). Internal (on-

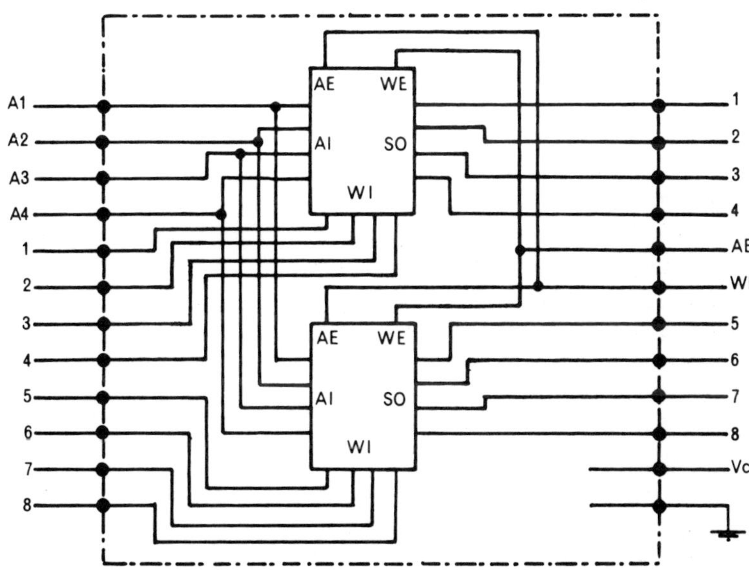

Figure 17.6. Arrangement for 16-word × 8-bit/word memory array.

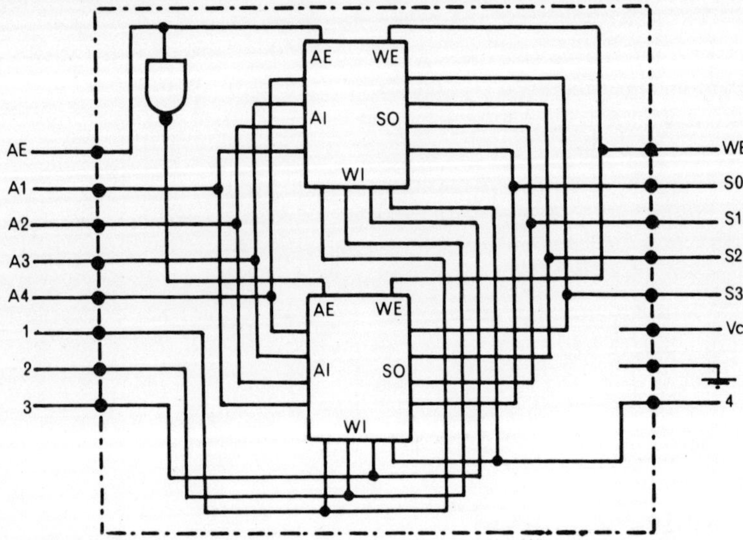

Figure 17.7. Arrangement for 32-word × 4-bit/word memory array.

substrate) decoding can be provided for the 32-word × 4-bit array; but since a total of only 17 substrate leads is required, and especially since only one single gate (for example, one-sixth of a hexagonal inverter package) is needed for the chip selection, it is rather expensive to mount this gate on the same substrate. Such a gate can be made available from the external logic.

17.6 A 256-BIT ARRAY

Four 64-bit memory elements are needed to form a 256-bit array, arranged as a 16-word × 16-bit block. They are wired together as a serial memory, with each word line consisting of 4 × 4 bit locations. Figure 17.8 shows such an arrangement, where the address-enable inputs and the address inputs and write-enable inputs, respectively, are paralleled together. Although this array is relatively simple, the total number of substrate contacts required is 40.

Four 64-bit memory elements are also required to enable the building of a 32-word × 8-bit memory array. Two ICs are paralleled together to form the 32-word lines and the eight bit locations of each word line. As shown in Figure 17.9, the sense outputs of each pair of elements are ORed together, as are the write-input group and the address-input group of each pair. A discrete inverting gate is used to select the first group of 16 word lines or the second one, as shown. This is also used to reduce

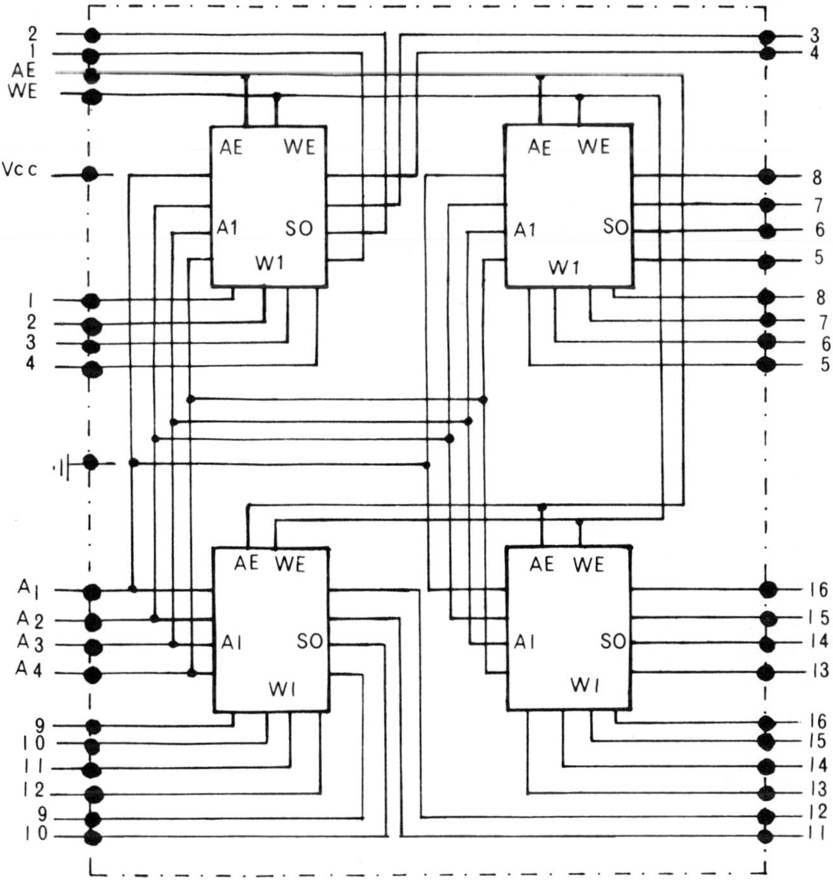

Figure 17.8. Arrangement for 16-word × 16-bit/word memory array.

the number of substrate connections to outside by one, making a total requirement of 24 leads. The additional decoding requires only one single gate; this gate must be provided on the substrate so that the maximum substrate leads does not exceed 24. This guarantees, in turn, that any package type can be used if it has 24 leads and is capable of housing the substrate.

The same number of four 64-bit memory cells can also be grouped together to provide a 64-word × 4-bit memory array as shown in Figure 17.10. Because the number of bits required is only four, all write-enable inputs, write-data inputs, and sense outputs are tied as shown. Also, the four address inputs can be paralleled together so that each group of 16 word lines can be selected by applying a logical "0" level to the appro-

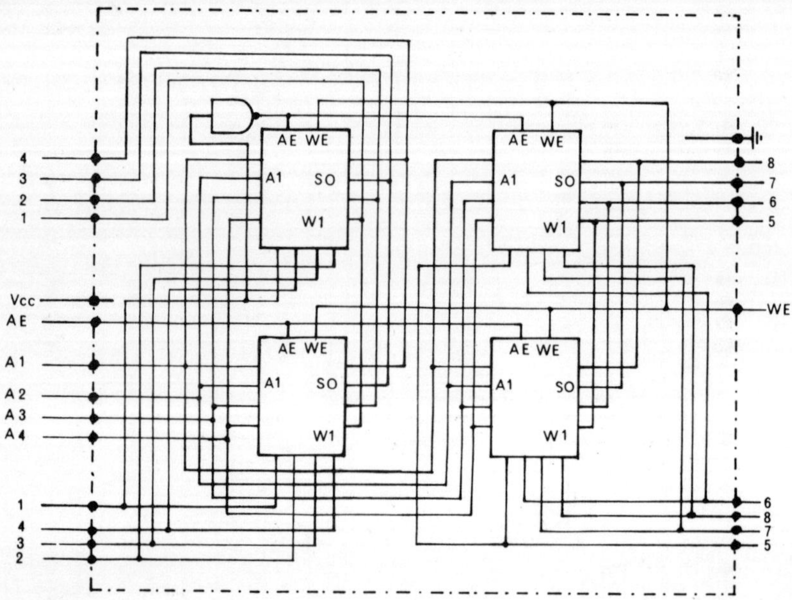

Figure 17.9. Arrangement for 32-word $\times$ 8-bit/word memory array.

priate address-enable input. The total number of substrate connections to outside can be reduced by two if a discrete 1-out-of-4 decoder network is provided on the same substrate using 2-quad two-input NAND gates. The total lead requirement is only 17.

17.7 A 512-BIT ARRAY

Eight 64-bit memory elements can be connected to make a 128-word $\times$ 4-bit array. All the sense outputs, write-enable inputs, address inputs, and write-data inputs are connected within each group. Each single address-enable input is a reference for each group of 16 word lines. Internal on-substrate decoding can be provided using a 1-out-of-8 decoding network consisting of a quad two-input NAND gate and three triple three-input NAND gates. The total substrate leads required will be reduced by five to only 18. Figure 17.11 shows a decoding network for the on-substrate construction of this array.

As shown in Figure 17.12, the eight 64-bit memory elements can be grouped together to provide a 64-word $\times$ 8-bit memory array. Four units are used to provide the 64 word lines, and the other four provide the eight

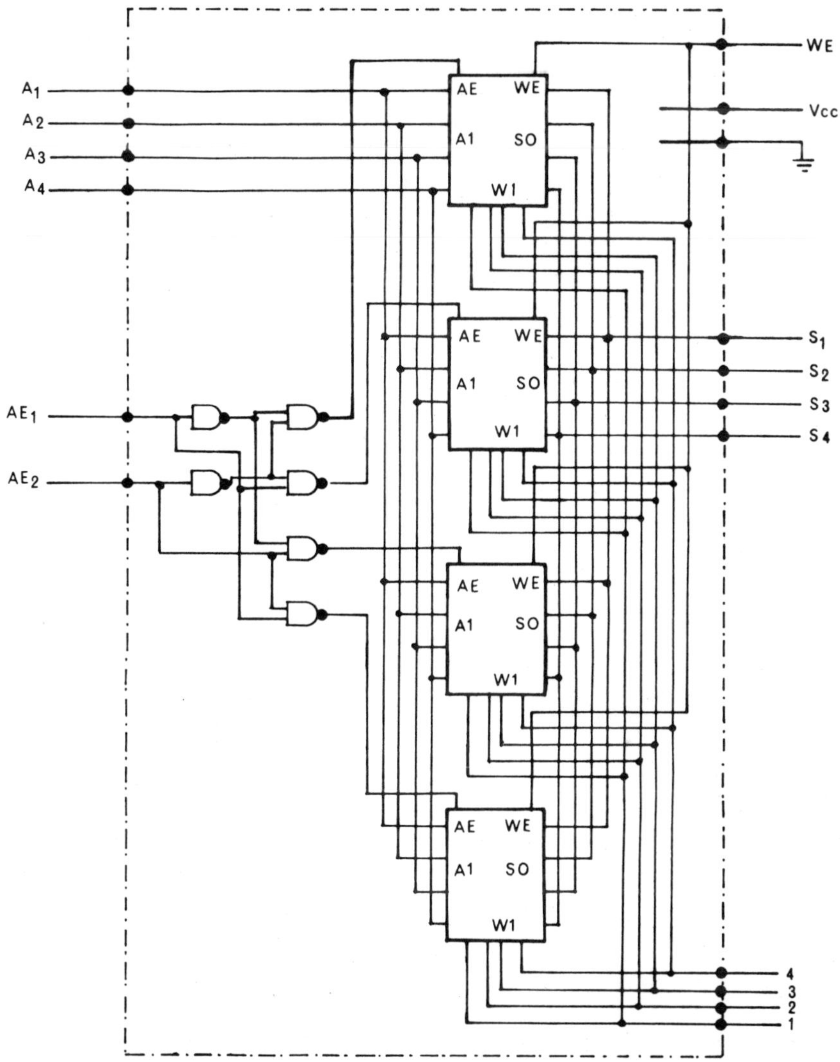

Figure 17.10. Arrangement for 64-word × 4-bit/word memory array.

bit locations for each word line. The sense output, write-enable inputs, address-enable inputs, write-data inputs, and address inputs are grouped together to form a matrix, with each matrix point representing a TMC6462/4IC. The total number of leads required by the substrate is 25 when a decoding network is also provided on the substrate. Using the same number of 64-bit memory elements as a building block, two other 512-bit array forms can be constructed a 32 words × 16 bits and 16 words ×

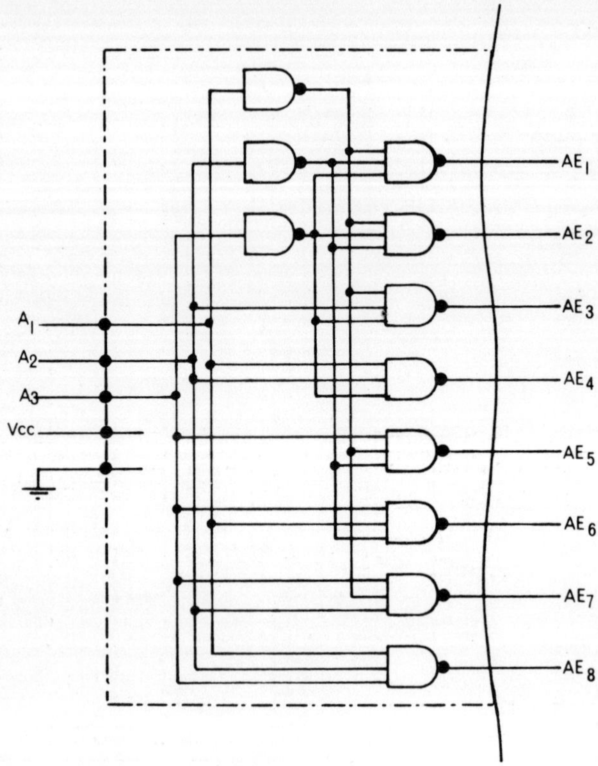

Figure 17.11. Decoding network for 128-word × 4-bit/word memory array.

32-bits. However, the leads required for the glass substrate increase in number to 40 and 72, respectively. This means that the handling of these modules becomes quite difficult and uneconomical.

17.8 ARRAYS OF 1024 BITS AND LARGER

By increasing the number of 64-bit memory elements to 16 or 32, memory arrays of larger capacities are possible. These include 1024 bits arranged as 16-word × 64-bit, 32-word × 32-bit, 64-word × 16-bit, 128-word × 8-bit, and 256-word × 4-bit arrays. Going one step further to 2048 bits, it is possible to construct 16-word × 128-bit, 32-word × 64-bit, 64-word × 32-bit, 128-word × 16-bit, 206-word × 8-bit, and 512-word × 4-bit memories. Of course, the number of leads required for the substrate and thus the package increases with the complexity of the arrangement. The presence of the address-input decoding network on the substrate is certainly helpful in reducing the total number of leads required

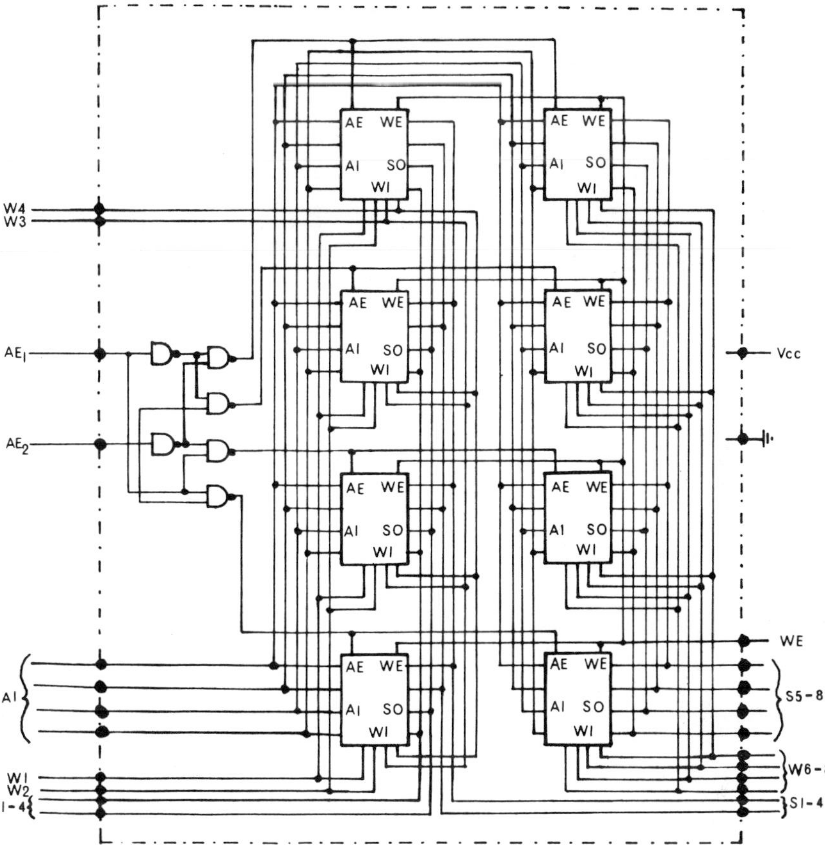

Figure 17.12. Eight 64-bit memory elements grouped together to form a 64-word × 8-bit/word memory array.

by this substrate. The lead reduction in some cases is quite substantial. To take an example, a 128-word × 8-bit memory hybrid array normally requires 31 connections to external logic; if on-substrate decoding is provided, however, this number is reduced to 26. Comparing a 64-bit memory cell, arranged as 16 words × 4 bits and a 1024-bit memory array, arranged as 128 words × 8-bits, both with internal decoding facilities, it is seen that the number of leads required by the package increases from 16 to 25, which is approximately 63%.

The relations between the total number of pins required for large memory arrays, with capacity expressed in bits, and possible organizational structures are depicted in Figure 17.13. These arrays would be built on glass substrates with metalization paths as wide as 10 mils and would in-

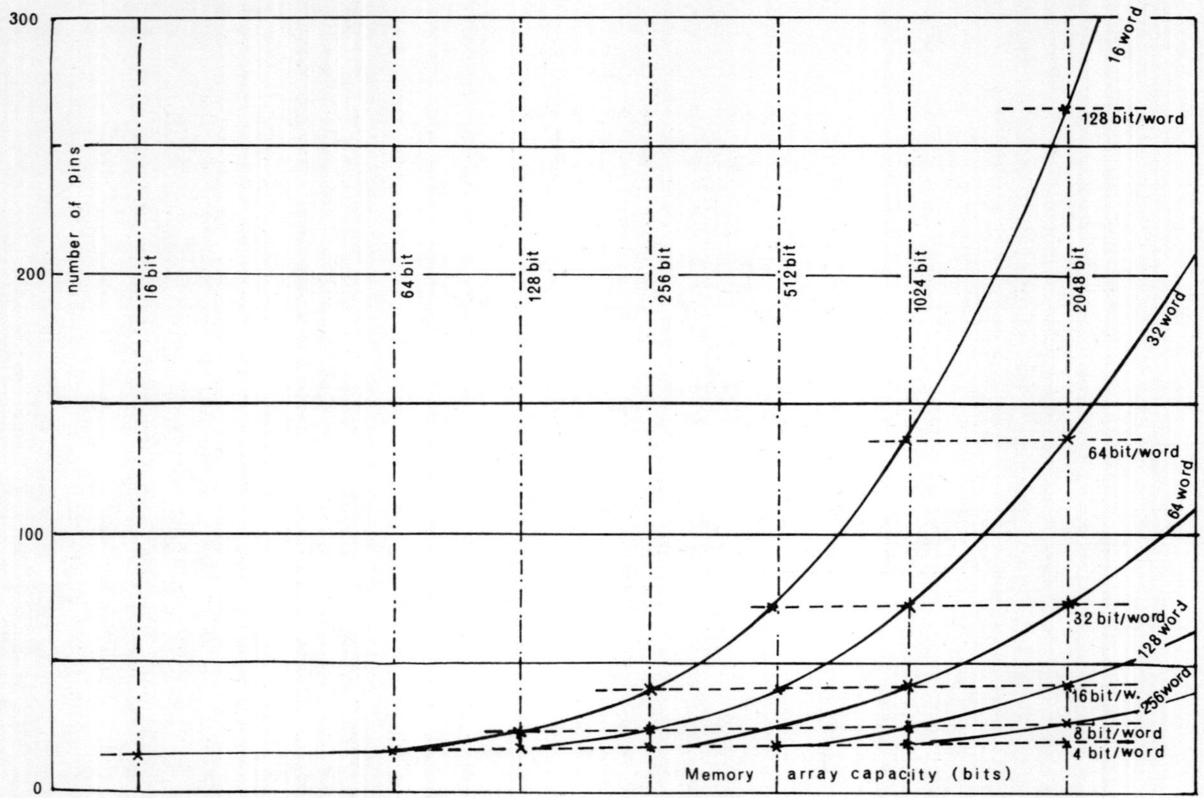

Figure 17.13. Relation between pin requirement, capacity, and organizational structure of large memory arrays.

TABLE 17.1 Bit Organization and Pin Requirements of Large Memory Arrays

Bit Capacity	Array Organization	Number of ICs (such as TMC6462/4)	On-Substrate Decoding	Pin Requirement	Substrate Area (in.)	Total Power Dissipation (mW)
128-bit	16 word × 8 bit	2		24	0.5 × 0.5	700
	32 word × 4 bit	2	¼ TNG3414	17	0.5 × 0.5	715
256-bit	16 word × 16 bit	4		40	0.8 × 0.8	1,400
	32 word × 8 bit	4	¼ TNG3414	24	0.8 × 0.8	1,415
	64 word × 4 bit	4	1½ TNG3414	17	0.8 × 0.8	1,490
512-bit	16 word × 32 bit	8		72	1.2 × 1.2	2,800
	32 word × 16 bit	8	¼ TNG3414	40	1.2 × 1.2	2,815
	64 word × 8 bit	8	1½ TNG3414	25	1.2 × 1.2	2,890
	128 word × 4 bit	8	1 TNG3414 + 2⅔ TNG3314	18	1.35 × 1.35	2,980
1024-bit	16 word × 64 bit	16		136	2.0 × 3.0	5,600
	32 word × 32 bit	16	¼ TNG3414	72	2.0 × 3.0	5,615
	64 word × 32 bit	16	1½ TNG3414	41	1.8 × 1.8	5,690
	128 word × 8 bit	16	1 TNG3414 + 2⅔ TNG3314	26	1.8 × 1.8	5,780
	256 word × 4 bit	16	1 TNG3414 + 8 TNG3114	19	1.8 × 1.8	5,900
2048-bit	16 word × 128 bit	32		264	3.5 × 4.5	11,200
	32 word × 64 bit	32	¼ TNG3414	136	3.5 × 4.5	11,215
	64 word × 32 bit	32	1½ TNG3414	73	3.5 × 3.5	11,290
	128 word × 16 bit	32	1 TNG3414 + 2⅔ TNG3314	42	3.5 × 3.5	11,380
	256 word × 8 bit	32	1 TNG3414 + 8 TNG3114	27	3.5 × 3.5	11,500
	512 word × 4 bit	32	1¼ TNG3414 + 32 TNG3014	20	3.5 × 3.5	11,755

corporate a decoding facility made out of discrete integrated circuits, to minimize the number of leads and, therefore, the package dimensions.

Figure 17.13 demonstrates that the rate at which the number of bits per word increases has a greater effect on the speed with which the number of pins required for hybrid large memory subsystems increases than does the rate at which the number of words increases. When designing such subsystems, therefore, efforts must be made to keep the number of contacts to the outside to a minimum for a given memory capacity. This is done by keeping the number of bit locations of each word line as low as possible.

As shown in Figure 17.13, memory arrays with any bit capacity and any bit per word organization can be achieved by combining a number of 64-bit memory elements. The total power dissipation of such a system is, however, quite high. Table 17.1 gives the typical power dissipation of various memory arrays with on-substrate decoding, as discussed above. It is clearly indicated that memory arrays that have a large bit capacity require a proper cooling facility. The high power dissipation suggest that the total system should be divided into small modules, each using a small number of 64-bit memory elements, arranged to provide the required memory organization.

17.9 CONCLUSIONS

Large memory capabilities are being achieved by combining tighter component tolerances, thinner epitaxial layers, shallower emitter diffusions, narrower isolations, and thinner metalization tracks, and using multimetalization. The 256-bit bipolar memory cell will be very competitive with MOS memory elements, both in component density and in power dissipation. An average power dissipation of 0.35 mW per bit will be possible, and at the same time speed will be five to 15 times as fast, which means that access time will be approximately 35 ns. Using 256-bit memory elements, much larger memory arrays, such as 4096- and 8192-bit, will be achieved with a reduction in power consumption of at least five to ten times.

18

Medium-Scale-Integration Definitions

by Robert Juan

Signetics Corporation

This chapter will supply the reader with definitions for several of the terms to be encountered when working in the MSI (medium-scale-integration) and LSI (large-scale-integration) areas.

18.1 SHIFT REGISTERS

A shift register is an arrangement of an arbitrary number of binaries in a row used primarily for temporary storage of digital information. A shift register whose length is n is capable of storing 2^n total states. The maximum magnitude of the stored decimal equivalent number is $2^n - 1$. An example of the storage capacity of a 3-bit shift register appears in Figure 18.1. This register is capable of storing 8 (2^3) total states, representing the decimal equivalents 0 through 7 ($2^3 - 1$).

18.2 SHIFT REGISTERS WITHOUT FEEDBACK

Common applications for shift registers include serial data entry and serial data output. In these applications, shown in Figure 18.2 and Figure 18.3, binary data are entered, one bit at a time. Each bit is entered with a clock pulse. Eight clock periods following the entry of the first bit and this

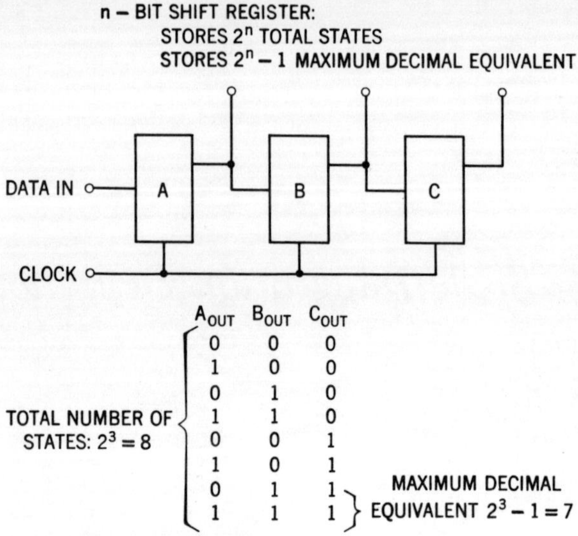

Figure 18.1. Three-bit shift register. R = clear; S = preset.

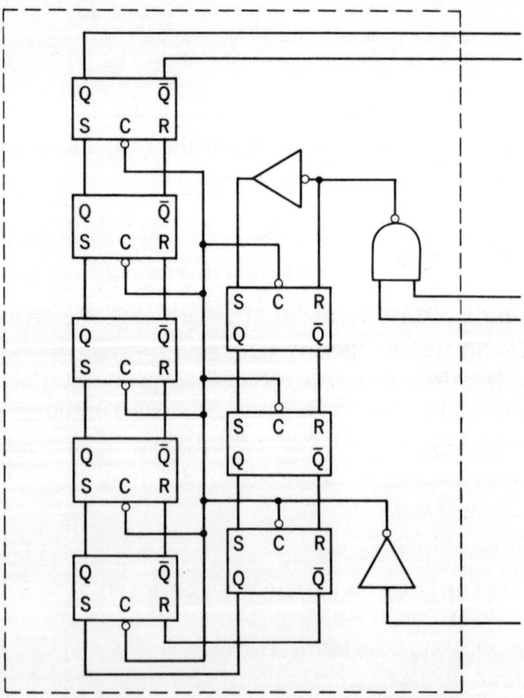

Figure 18.2. Eight-bit shift register.

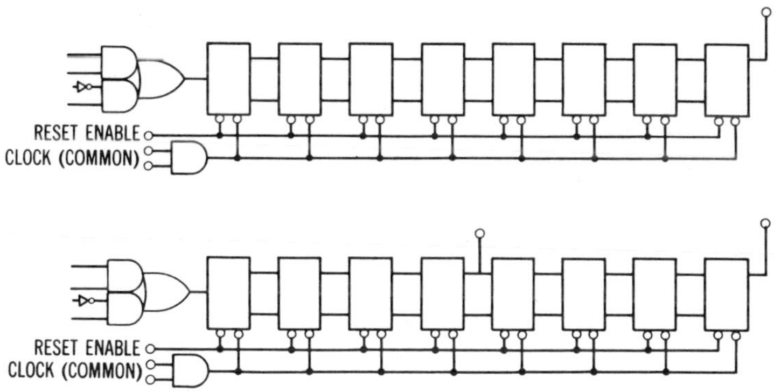

Figure 18.3. Dual eight-bit serial shift register.

binary information is stored at the serial output. Thus each bit of data is delayed between the input and the output by eight clock cycles. It is not surprising that most serial-in serial-out shift registers are used as delay lines in digital systems.

A second common shift-register arrangement is shown in Figure 18.4. This is a serial-to-parallel converter. Each bit of data is entered with a clock pulse. After nine activations of the clock input, an entire 8-bit word with its parity (sign) indication is stored and ready for parallel transfer or operation. A popular application for this type of device is an accumulator.

An accumulator is used where data are entered in serial fashion (as from a keyboard) and then acted upon in parallel (as by an adder). The parity bit allows expansion of the add function to include subtraction, multiplication, and division. Parallel-to-serial converters as shown in Figure 18.5 complement serial-to-parallel devices. In general, this form of shift register is used to accumulate binary words from a parallel operation, such as addition, and feed them serially into another storage area, such as a display or a memory.

In complex arithmetic sections it is often desirable to be able to serially shift information contained in a register in either direction. An example of

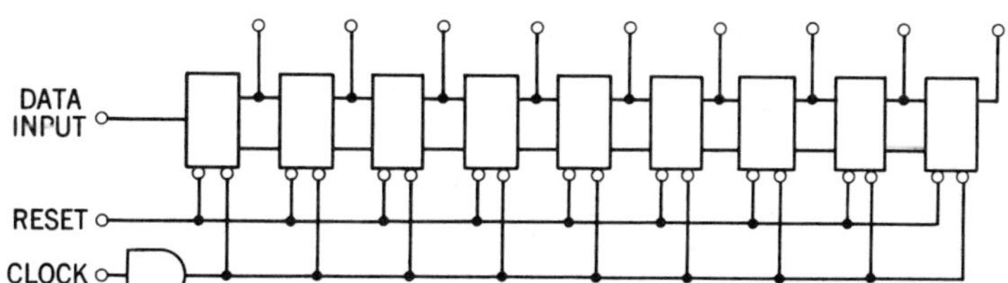

Figure 18.4. Nine-bit serial-to-parallel converter.

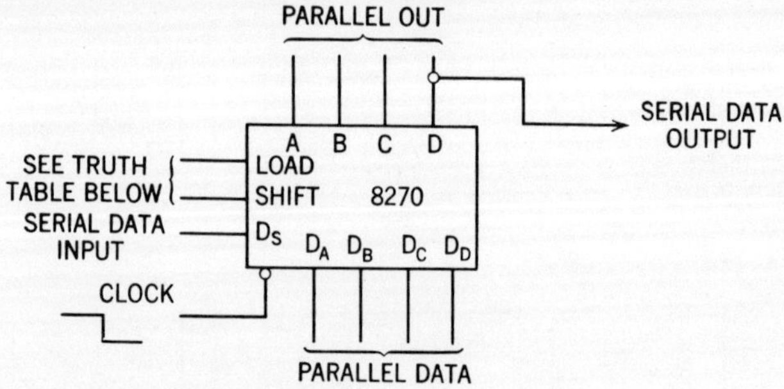

Shift	Load	Function at Clock Fall
0	0	Storage
1	0	Shift right (serial)
0	1	Load (parallel)
1	1	Shift right (serial)

Figure 18.5. Parallel-to-serial converter.

a bidirectional serial shift register of shift-left, shift-right serial shift register is shown in Figure 18.6. Besides serial shift in either direction, some applications require the ability to accumulate a bit of data from a parallel operation and then shift the word either to memory or to display, depending on the next command. This can be accomplished with the bidirectional serial shift register with parallel entry shown in Figure 18.7.

18.3 SHIFT-REGISTER APPLICATIONS WITH FEEDBACK

In addition to simply storing digitial information, certain unique functions can be generated using shift registers with feedback. Each binary in a

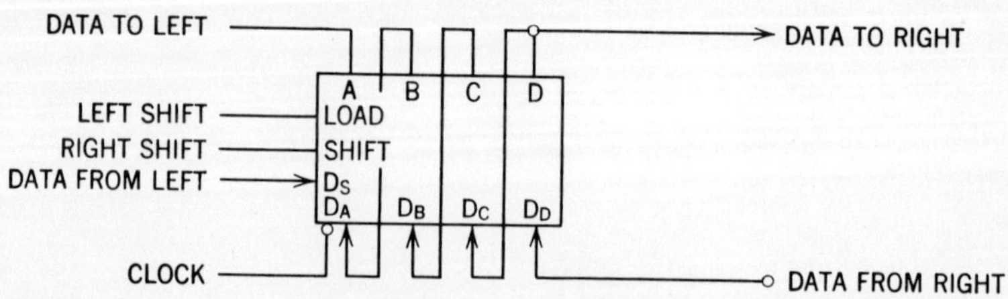

Shift	Load	Function at Clock Fall
0	0	Data stored
0	1	Shift left
1	0	Shift right
1	1	Shift right

Figure 18.6. Bidirectional serial shift register.

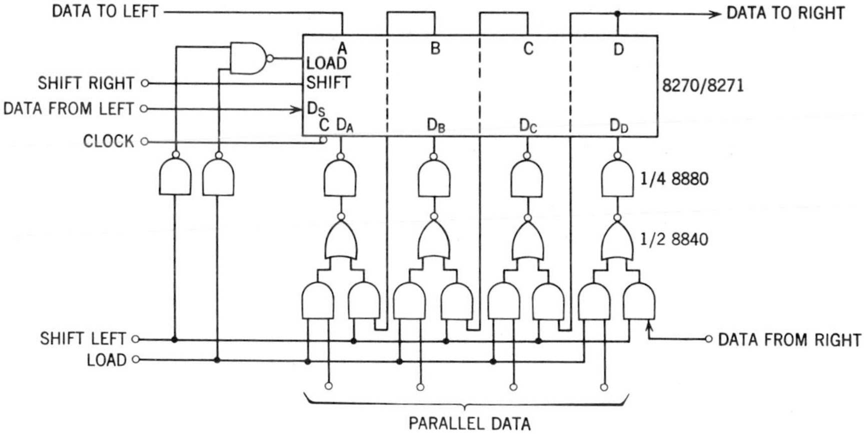

Figure 18.7. Bidirectional serial shift register with parallel entry.

shift register transfers its stored data to the adjacent binary each time a clock pulse occurs. An n-bit shift register will be inactive (all outputs equal logical "0") after n clock pulses unless new logical "1" signals are introduced at the first bit. One way to keep a register active is to feed back the states of certain binaries to the input binary.

18.4 MAXIMUM-LENGTH SEQUENCE GENERATORS

By proper choice of feedback and logic and the proper selection of the bits to be fed back, maximum-length sequence generators can be constructed. The maximum length of the sequence generated by a feedback shift register of n binaries is equal to $2^n - 1$. Figure 18.8 shows a 15-state maximum sequence generator incorporating a 4-bit shift register and one-fourth of a quad-exclusive OR gate. Zero suppression may be added to

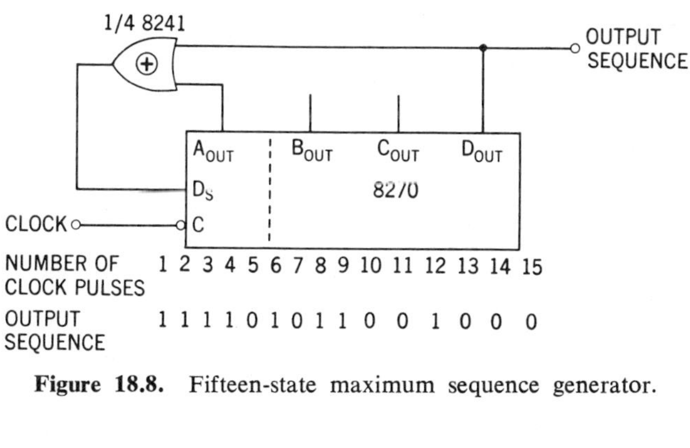

Figure 18.8. Fifteen-state maximum sequence generator.

ensure that the register remains active in the event the initial state of all binaries equals "0." Maximum-state sequence generators find wide application in generating specially shaped pulse trains.

18.5 SEQUENCE GENERATOR WITH PHASE-RELATED OUTPUTS

Another version of the maximum sequence generator that finds wide application in system timing circuits is shown in Figure 18.9. Here, the two outputs are phase related. Figure 18.9 specifically demonstrates a 7-state maximum sequence generator where the sequence generated at output number 2 leads an identical sequence generated at output number 1 by three clock pulses.

18.6 ENCODING AND DECODING BINARY WORDS

In transmitting digital data serially, a need may arise to encode the transmitted message and then decode the data received at another location, retrieving the original message.

Figure 18.10 depicts such a transmitter and receiver. In this example, a binary word, modulo 2, is added to the output sequence of the transmitter. Decipherment consists of adding the same key, modulo 2, to the encoded message using a second generator with an identical synchronous sequence. Both generators used in encoding and receiving must have the same se-

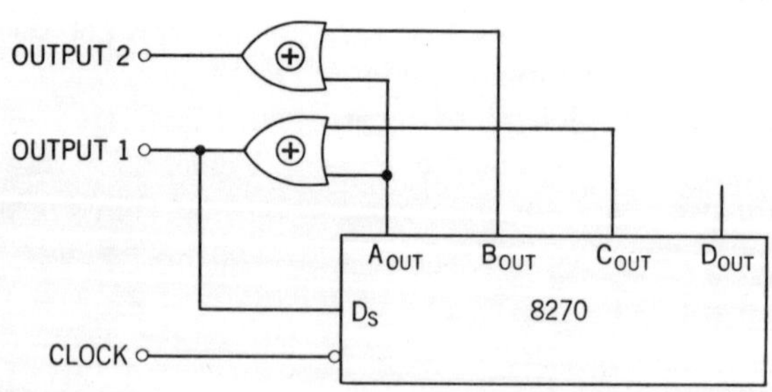

CLOCK PULSE	OUTPUT 1	OUTPUT 2
1	1	0
2	1	1
3	1	0
4	0	0
5	1	1
6	0	1
7	0	1
8	1	0

Figure 18.9. Seven-state sequence generator with phase-related outputs.

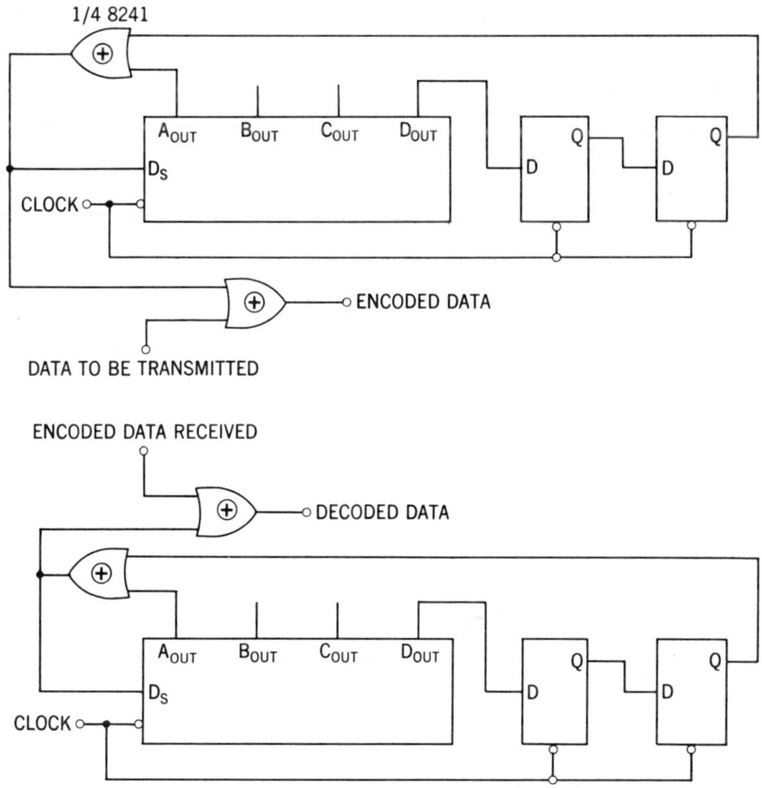

Figure 18.10. Transmitter and receiver for encoding and decoding digital data.

quence and must be clocked in phase. Using the pseudo-random feature of a much longer shift register will result in a more powerful encipherment.

18.7 VARIABLE-FREQUENCY DIVIDER

A modification of the basic maximum sequence generator can be designed to operate as a variable-frequency divider. This is accomplished by detecting all the "1" states of the n-bit register and then parallel loading a particular n-bit word. The parallel loading of a particular word shortens the maximum sequence inherent in the register, resulting in a frequency divider in which the output switches at a specific fraction of the input clock frequency.

The total number of variable output frequencies for an n-bit shift register ranges from $f/2$ to $f/2^n - 1$, where f is the input clock frequency. Figure 18.11 shows a variable-frequency divider of eight total bits. It is capable of

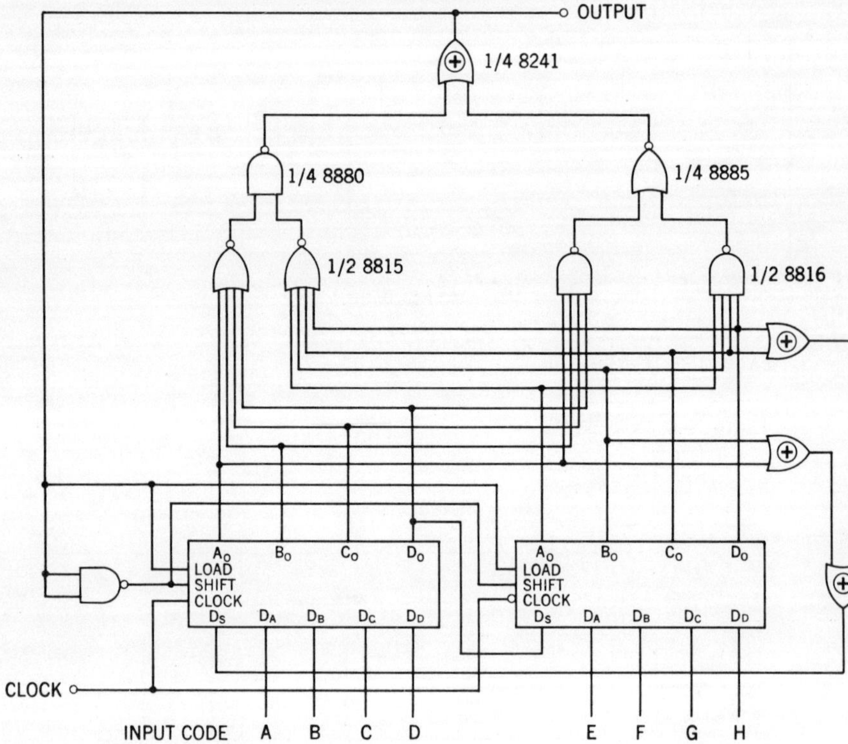

Figure 18.11. A variable-frequency divider of eight total bits.

dividing the input clock frequency by every integer from 2 to 255 ($2^8 - 1$). The frequency division is controlled by the 8-bit input code.

19

Converting from Binary to Binary Coded Decimal Numbers

by Edward F. Tarbox

Sylvania Electric Products, Inc.

Bipolar read-only memories provide a fast and easy way to convert binary numbers to binary coded decimal (BCD) numbers; all that is really needed is a decoder. A ROM is well suited for this type of work.

In describing the conversion, an SM320 ROM will be used. This IC consists of a 1-out-of-32 decoder addressed by a 5-bit binary number, a decode or chip enable, and 32 eight-bit storage locations. The 1-out-of-32 decoder selects one of the 32 eight-bit storage locations, which in turn presents the bit pattern stored at that location to the output. The eight outputs have free collectors to permit expansion of the size of the memory by wire ORing and and ANDing the outputs of two or more ROMs with external pull-up resistors.

By making use of the chip enable and connecting the outputs of two ROMs in parallel, a memory of 64 eight-bit words can be constructed; or by expanding even further, four ROMs can be used to construct a 128-word, 8-bit memory. These memories can also be expanded in the other direction. That is, two ROMs can be used to make up a 32-word, 16-bit memory. For a binary-to-BCD converter, it is only necessary that the stored bit pattern of any given location be the BCD value of the binary address for that location. Four of the outputs are used for the "units" value of the BCD number or character and four of the outputs are used for the "tens" value of the BCD character. Therefore, to provide enough memory to convert a binary number to a two-character BCD number, 0 through 99, it will be necessary to have 100 storage locations of eight bits each. This requires four ROMs connected as a 128-word, 8-bit memory.

Thus, the first ROM will contain the BCD values of 0 to 31; the second ROM, the BCD values 32 to 63; the third ROM, the BCD values 64 to 95; and a fourth ROM will contain the BCD values 96 to 99. The sixth and seventh bits of the binary number address a 1-out-of-4 decoder, which selects the ROM that is active. A block diagram of this method is shown in Figure 19.1.

A closer look at the problem will allow some simplification. Since every odd value of a binary number is also odd in its BCD equivalent (and the same is true of even numbers), it is necessary to decode the least significant bit of the binary number. If this bit is brought out and becomes the least significant bit of the BCD number, it is necessary only to decode even binary numbers. Therefore, the only storage needed is for 50 BCD numbers with seven bits per word. This requires two ROMs with the first three outputs used for the "units" BCD character, the least significant bit of the binary number is the fourth output, and the next four outputs are used for the "tens" BCD character. With this method, the eighth output can be used for the least significant bit of the "hundreds" BCD character, which in turn

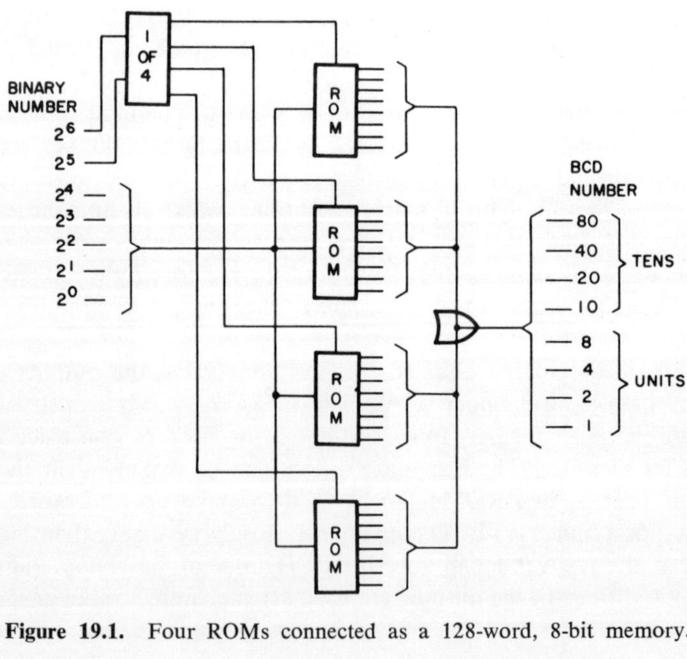

Figure 19.1. Four ROMs connected as a 128-word, 8-bit memory.

allows for decoding the full seven binary bits from 0 to 127. The layout for this method is shown in Figure 19.2. This technique only requires two ROMs, and an inverter to handle the chip enable, and is therefore a significant improvement over the arrangement shown in Figure 19.1.

19.1 HANDLING LARGER BINARY NUMBERS

The same procedure may be used to handle larger binary numbers for conversion to three or four decimal characters, but the number of ROMS required will increase rapidly. For example, to handle a three-character BCD number 000 to 999, at least 500 storage locations will be needed; a four-character BCD number would require 5000 locations. Inasmuch as these BCD numbers would require more than seven outputs, four outputs per BCD character, the memory size would have to be 500 or 5000 by 16 bits.

Although this is highly impractical, a system has been developed that takes full advantage of the basic 7-bit binary-to-BCD decoder to handle larger numbers. As an example, we'll consider the conversion of a 13-bit binary number to a four-character BCD number.

The seven lower-order bits of the 13-bit binary number are converted to a BCD number using the method described previously. The next six higher bits are then used as inputs for a second ROM decoder. This memory is arranged to have sixty-four 16-bit locations. The 64 locations are required

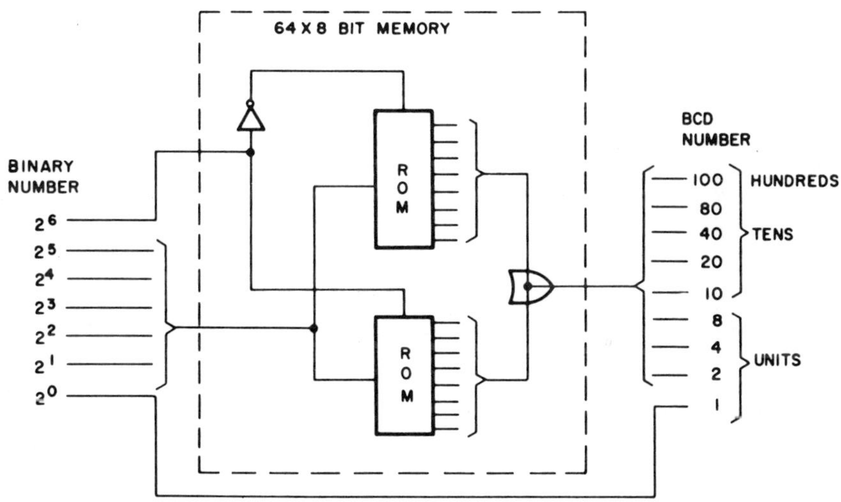

Figure 19.2. Two ROMs combined with an inverter provide a significant improvement over the arrangement shown in Figure 19.1.

to handle the six binary numbers, since the maximum value of a 6-bit binary number is 64. The 16-bit word length is required because the possibility exists of four BCD characters from the six higher binary numbers, and each BCD character requires four bits. Four bits are required for the units, four bits for the tens, etc.

The bit pattern stored in each location of this memory is then the BCD value of the binary address of bits 8 to 13. Thus, the first location in this memory will contain all zeros, the second location will contain the BCD number 128, the third location will contain BCD number 256, the fourth location will contain the BCD number 384, etc. The location contents will continue to increment by 128 up to the last location, which will contain the BCD number 8064.

To accomplish this, the five bits 2^7 through 2^{11} are used as the address of the four ROMs in the 64-word, 16-bit memory. The thirteenth, or 2^{12}, bit is used as the chip enable to select which memories are to be used. For example, if bit 2^7 is true and bits 2^8 through 2^{12} are false, the output of the memory will be 0000, 0001, 0010, and 1000, which is 128 in BCD notation—the value of bit 2^7. If all bits 2^7 throug 2^{12} were true, the output of the memory would be 1000, 0000, 0110, and 0100, which is 8064 in BCD notation—the value of bits 2^7 through 2^{12}.

The next step is to take the output of this memory and add it to the BCD value obtained from decoding the first seven bits. If all 13 bits are true, the first seven bits will give use 127 which, when added to the 8064 from the six higher bits, will give us 8191. Figure 19.3 is a block diagram of this procedure.

19.2 PERFORMING ADDITION

Before examining binary numbers of other sizes, a way of doing the addition should be considered. Since the addition that must be performed involves BCD numbers, a binary adder cannot be used. However, one way around this is to convert the BCD numbers to excess-3 notation, add with binary adders, and then convert the result from excess 6 to normal BCD where required. The process of converting the numbers to excess 3 can be eliminated by having the outputs of the 64-word, 16-bit memory, used for values greater than 127, expressed in excess-6 BCD notation. When the two numbers are added, this is equivalent to converting both numbers to excess 3 and then adding. This procedure is shown in Table 19.1 for the value 8191.

To convert the result of the additions to BCD, the sum of each decade is treated separately in the following manner: If the carryout of the decade is at "1," then no further action is required (or, in this case, 0 is subtracted

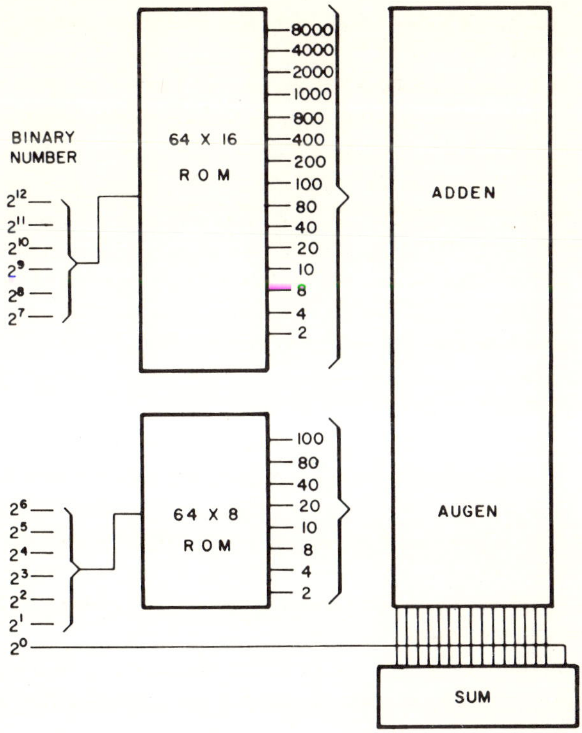

Figure 19.3. Conversion of 13-bit binary number to four-character BCD number using 64-word, 16-bit and 64-word, 8-bit ROM memories.

TABLE 19.1 Example of Conversion to Excess-6 BCD Notation

Output from bits 1 through 7:		0000	0001	0010	0111	= 127
		1000	0000	0110	0100	= 8064
		0110	0110	0110	0110	+ 6 for each decade
Output of bits 8 through 13:		1110	0110	1100	1010	= 8064 (×6)
	0	0	0	1		
		1110	0110	1100	1010	= 8064 (×6)
		0000	0001	0010	0111	= 127
Result of additions		1110	0111	1111	0001	= 14 7 15 1
Conversion from × 6 where required		1110	0111	1111	0001	
		1010	1010	1010	0000	
		1000	0001	1001	0001	8191

from the sum of the decade). If the carryout of the decade is a "0," then 6 is subtracted from the sum of the decade. Subtracting 6 from the decade, where necessary, is accomplished by adding a —6 expressed in 2's complement and ignoring any resultant carryout that may be generated. In this case, only the units character had a "1" for a carryout, so 0 is subtracted from this decade and —6 is subtracted from the three other decades.

Figure 19.4 shows the logical necessary to convert a 13-bit binary number to a four-character BCD number. The adders *L* and *M* are 3-bit adders because the —6, or —0 does not involve the least significant bit of a BCD character. Also, the BCD values for the thousands and hundreds characters do not have to be expressed in excess 6, since there is room in adders *I* and *J* to add 6 to each value directly. Also, since we are dealing with a maximum value of 8191, the thousands BCD character can never have a value greater than 8, and thus binary adders are sufficient for this decade, eliminating adder *N*. The values stored in memories *C* and *D* must be in excess-6 BCD notation, for there is no room to add 6 separately to the units and tens decade. Therefore, to performing the 13-bit-binary–four-character-BCD conversion take six ROMs and 24 full adders. Any inverters necessary to implement this system can be obtained from SM10 full adders, which also contain an independent three-input NAND gate.

If 10-bit-binary–three-character-BCD conversion is all that is required, the only requirements are that the second decoder be capable of handling

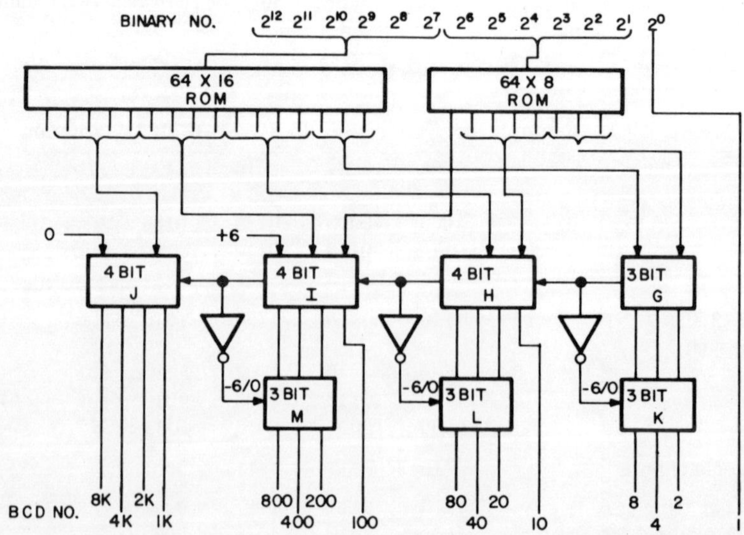

Figure 19.4. Logic required for converting a 13-bit binary number to a four-character BCD number.

a three-bit address and that a 32-word, 16-bit memory can be used. This method, shown in Figure 19.5 will allow full decoding to 1023, the maximum value of a 10-bit binary number. The conversion requires four ROMs and 20 full adders.

19.3 DECODING METHODS

If it is desired to decode to a full four-character BCD number from 14 binary bits, which actually has a maximum value of 16,384, there are two approaches to consider. One is to add the value 8192, the value of the fourteenth bit, to the results of converting a 13-bit binary number to a BCD number. The value of the fourteenth bit, 2^{13}, expressed in excess-6 notation, can be wired into a 15-bit adder along with the results from Figure 19.4. Twelve additional adders will have to be used to convert the resulting sum to BCD notation. This method takes an addition of 27 full adders to add the fourteenth bit to the results of the 13-bit converter, which took 30 units, for a total of 57 units (see Figure 19.6).

The second approach is to expand the 64-word, 16-bit memory to a 128-word, 16-bit memory. This requires the addition of four ROMs plus a 1-out-of-4 decoder for the chip enable. This method requires a package count of 10 ROMs, a 1-out-of-4 decoder, and 27 adders, for a total of 38 units. From the standpoint of package count, this method is definitely superior.

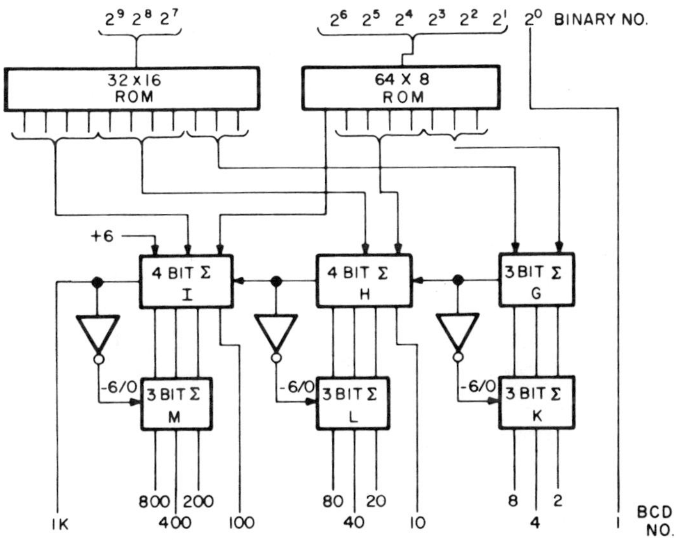

Figure 19.5. Arrangement for full decoding to 123, the maximum value of a 10-bit binary number.

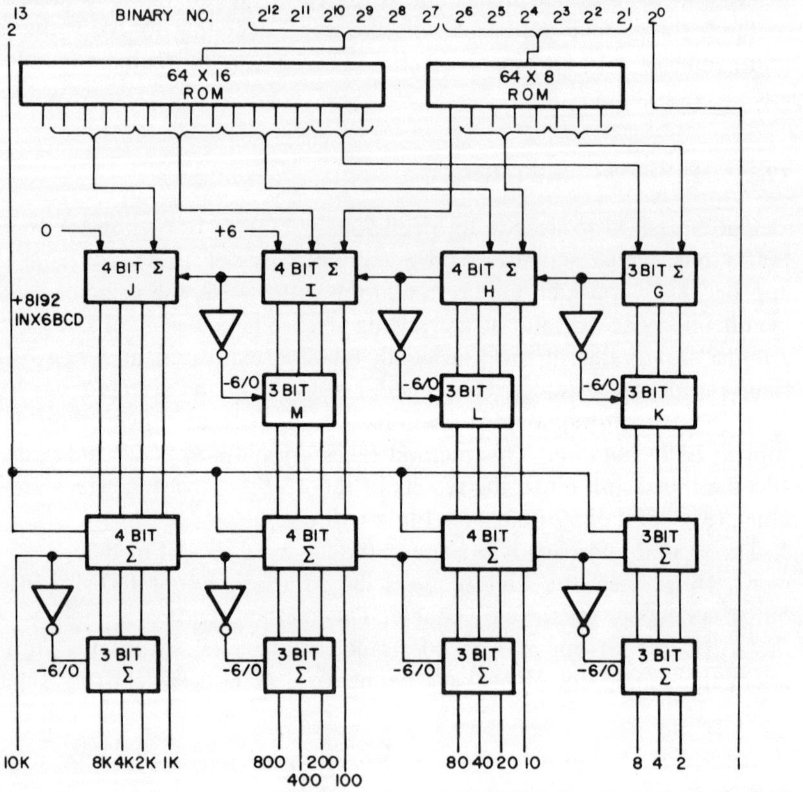

Figure 19.6. Arrangement for decoding a full four-character BCD number from 14 binary bits (maximum value of 16,384).

However, if it is assumed that the ROMs cost about ten times as much as adders, then a cost weighting factor can be applied to each method. This method will have a cost weighting of $(10 \times 10) + (1 \times 28)$, or 128, for 38 units, whereas the first method—just adding adders—will have a weight of $(6 \times 10) + (1 \times 51)$, for a total of 111 for 57 units (see Figure 19.7). The weighting factor of 10 is purely arbitrary; it is included only to facilitate a comparison of the methods and does not reflect the actual cost of any supplier's devices.

To expand the system to handle 16 binary bits, there are, as before, two approaches available. One is to expand the second memory to a 512-word, 16-bit array, which requires 32 ROMs plus some gating to generate the fifth decimal character. This would give a cost weighting of $(10 \times 34) + (1 \times 35)$, for a value of 375 for 69 units (see Figure 19.8). The other approach would be to have bits 14, 15, and 16 address three ROMs and

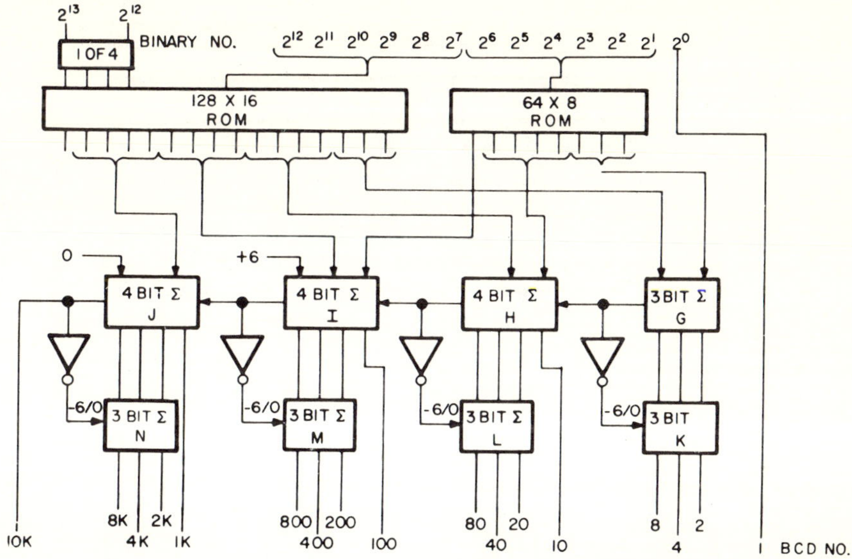

Figure 19.7. Alternate setup for arrangement shown in Figure 19.6.

generate eight 5-BCD characters, which can then be added to the results obtained from decoding the first 13 bits. This will require three more ROMs and 30 more adders. This results in a cost weighting of $(10 \times 9) + (1 \times 54)$ or 144 for 63 units (see Figure 19.9).

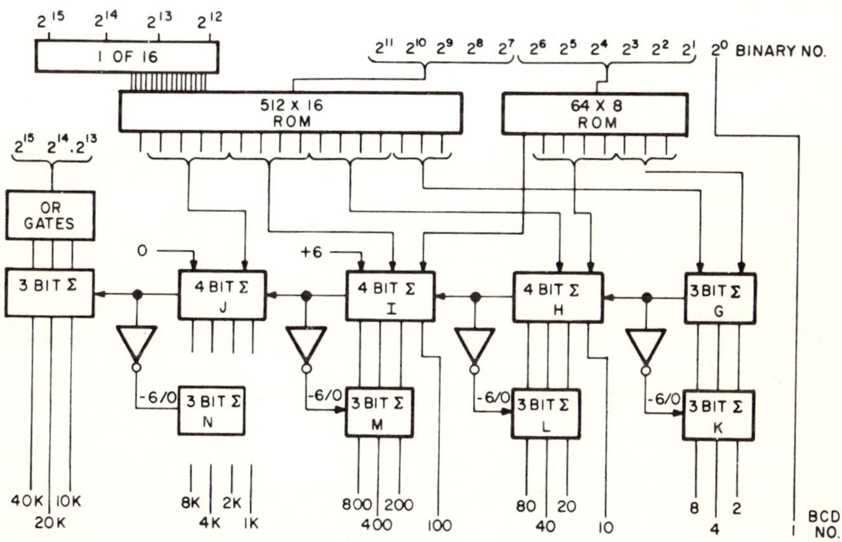

Figure 19.8. Expansion of system to handle 16 binary bits.

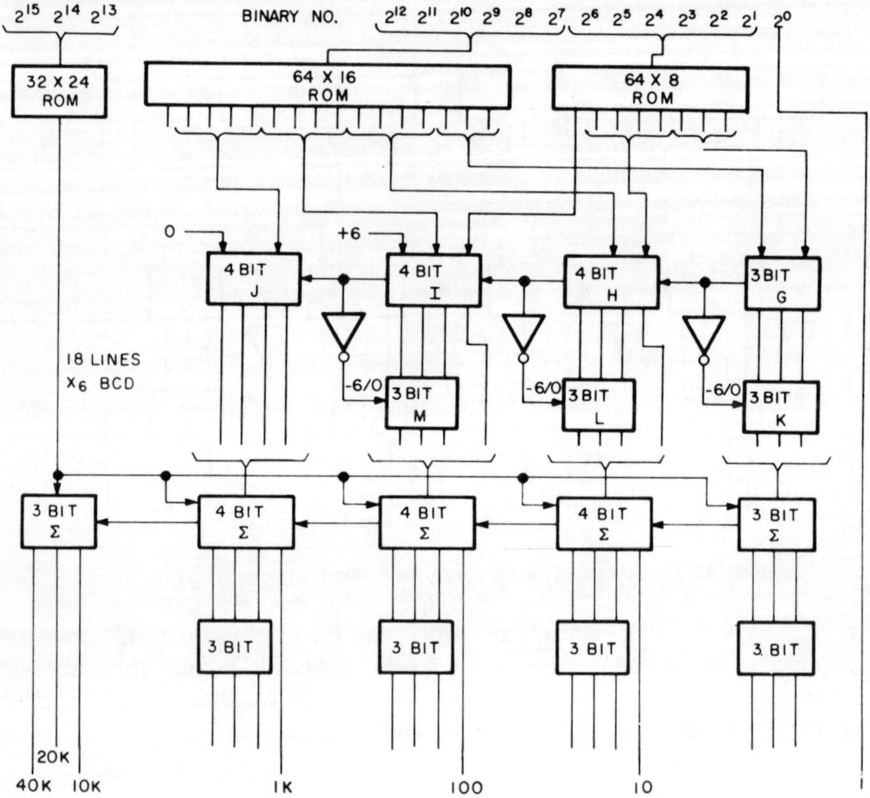

Figure 19.9. Alternate approach to arrangement shown in Figure 19.8.

Table 19.2 summarizes the unit count and cost weighting for the number of conversions discussed. These methods of conversion may seem to take

TABLE 19.2 Unit Count and Cost Weighting

Number of binary bits	7	10	13	14	14	16	16
Number of ROMs	2	4	6	10	6	34	9
Number of adders	0	20	24	28	51	35	54
Unit count	2	24	30	38	57	69	63
Cost weight	20	60	84	128	111	375	144

more units to do the job and to be more costly than the conventional shifting techniques, but they offer such advantages as high speed, simple circuit design, clockless operation, and minimum number of special MSI or LSI circuits.

Whenever an additional stage of arithmetic addition is required, the conversion time increases; however, increasing the size of the memories does not appreciably increase the conversion time. If high speed is a definite requirement, then anticipated carry adders can be used. Figure 19.10 is a graph of conversion times for the number of bits used, both for ripple adders and look-ahead adders. These conversion times are typical for SM10, 20, 30, and 40 series adders. Points are also indicated for a conversion by increasing the size of the second memory.

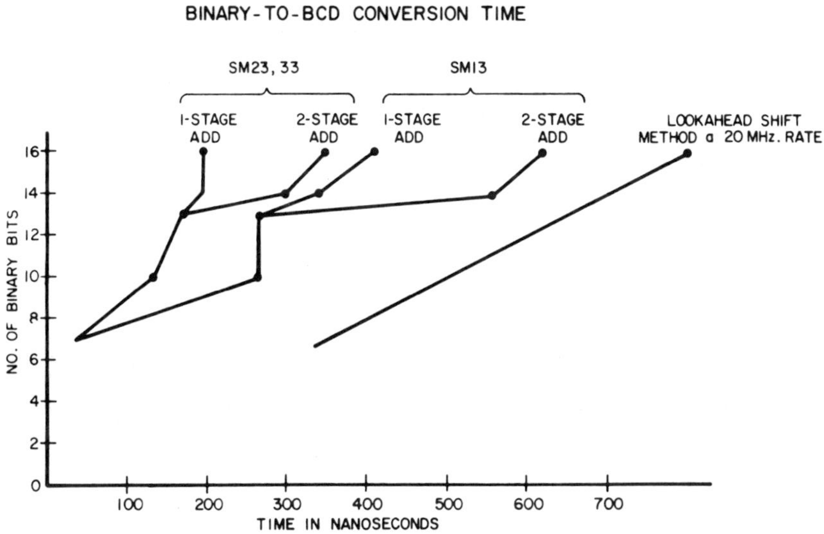

Figure 19.10. Conversion times for the number of bits used (for ripple adders and look-ahead adders).

20

Memory Cost and Reliability Considerations for Beam-Lead Memories

by Anthony Holbrook

Computer Microtechnology, Inc.

The criteria leading to an "optimum" memory choice should include reliability as well as cost and performance.

The reliability of any semiconductor product is influenced as much by the package in which it comes as by the process used to make it. The reliability of a mature semiconductor product fabricated by stable process technologies, therefore, can be adversely affected by the packaging technique employed and, specifically, by the number and types of interconnections. Since packaging is a prime determinant of product cost, any discussion of cost and reliability considerations obviously must include packaging scheme as well as semiconductor process.

For several years it has been argued that the ideal method for packaging ICs should involve beam leads. Since this belief has been carried over to memory ICs, the relation between reliability and the beam-lead approach will be studied in this chapter.

20.1 IC COST CONSIDERATIONS

Establishing an absolute cost for any semiconductor product is a complicated procedure involving the "yields" for many processing operations, the cost of raw materials, and the labor needed for each operation. Rather than trying to determine the absolute cost of one or more memory subsystems by suffering through many assumptions in an attempt to establish a series of individual costs, we'll discuss only the nature of each major cost factor. The following three areas will be discussed: integrated-circuit die cost, cost of assembling dies into packages, and cost of assembly of the memory system.

20.2 INTEGRATED-CIRCUIT DIE COST

The semiconductor chips used in both single-die-per-package systems and multichip modules are fabricated identically through the deposition of a metal interconnection pattern on the die. The IC die to be multichipped must have the additional processing step of depositing and defining the beam leads.

It is essential that the beam-lead process be simple and straightforward and result in a high yield of dies after deposition. The aluminum beam-lead process qualifies in all respects. It is a one-step process in which the beams are deposited by evaporating aluminum through a mechanical mask onto the die. The process maintains a single metal interconnection system and sidesteps problems associated with interfacing dissimilar metals. Because the boundaries of the beams are defined by the mechanical mask, there is no need for any etching or lifting process in order to achieve beam definition. The beams can be interdigitated between adjacent dies so that the beams of one die extend over the die next to it.

Additional space separating the dies on the wafer is not required by the beams, so the gross number of dies on a wafer is the same as it would be if the beam-lead process had not been used. Dicing of the wafer is performed after the beams are deposited.

Because of its uncomplicated nature and because it is a batch process, beam-lead deposition is economical. Dies intended for use in a conventional single-die-per-package system will cost less than beam-leaded dies, but with the aluminum beam-lead process, the difference in cost is estimated at less than 10%.

20.3 ASSEMBLY OF DIES INTO PACKAGES

A large memory system today involves the use of several IC chips in a package. A typical hybrid package will have: (1) layers of aluminum interconnections separated by dielectric and deposited onto a ceramic (alumina) substrate, (2) leads attached to the substrate, (3) IC chips bonded to the aluminum interconnection pattern, and (4) a lid sealed to the substrate, hermetically enclosing the IC chips.

20.4 MEMORY COST CONSIDERATIONS

The first major cost consideration for the hybrid memory package is the multilayer interconnection pattern on the ceramic substrate. The possible necessity for multilayer metal on the IC die can be avoided by using two layers of interconnection on the substrate. Because the interconnection is deposited on the substrate, the masking and process tolerances associated with the deposition can be large, resulting in a high-yield, low-cost process. In addition, a faulty interconnection on the substrate affects only the price of the raw substrate and the labor needed to deposit the interconnection. On the other hand, a faulty interconnection an an IC die with multilayer metalization results in the loss of a die that has been through numerous processing operations.

The second major cost consideration concerns the attachment and bonding of dies to packages. Any defect that occurs during the die attachment and bonding operation of a single-die-per-package results in the loss of that package and die. The cost of both must be amortized over the cost of good products made with the same operations. In a multichip beam-lead module, die attachment and bonding are performed in the single operation of bonding the beam leads to the aluminum metalization. Alignment of the beams to the metalization is done quickly and efficiently, without the need for sophisticated optical lineup systems. The beams are external to the periphery of the die, are clearly visible, and can be aligned directly to the interconnection pattern.

When a faulty beam-to-metalization bond does occur, two repair techniques can be used. First, the beam may be rebonded by simply reapplying the bonding tool to the beam. If that fails, the die may be removed and replaced; shearing the die from the substrate metalization removes the die without damaging the metalization. It is then possible to bond another die to the same location. The repairability aspect of aluminum-beam circuits makes them a prime consideration for high-reliability designs.

20.5 ASSEMBLY OF THE MEMORY SYSTEM

Building a memory system from multichip modules rather than from single-die packages results in a significant reduction in cost. A single-die-per-package subsystem equivalent in terms of performance and complexity to a 4096-bit module would be composed of 22 packages: sixteen 24-lead packages containing MOS chips, and six 16-lead packages containing bipolar chips. The 22 packages must be inserted into a multilayer printed circuit board. A hybrid module that directly replaces the 22 packages and their printed circuit board can be put in a 26-lead, dual-in-line 1.5-in^2 package.

Hybrid memory packages provide the system designer with: (1) considerably fewer printed circuit boards, (2) relaxed design constraints on their printed circuit board can be put in a 26-lead, dual-in-line 1.5-in^2 the printed circuit boards because of the larger ratio of package area to number of leads of the module package, (3) fewer insertions of components into each printed circuit board and into each system, (4) a space saving in the memory system because of the higher degree of integration, and (5) a decreased number of part types that must be kept in inventory.

20.6 RELIABILITY CONSIDERATIONS

The major sources of failure in integrated circuits are the bonds between die and package. The increased reliability of the beam-lead multichip module over the single-die-per-package system is dramatic. Table 20.1 shows failure rates published by Rome, New York, Air Development Center for several types of interconnections associated with both assembly techniques. Table 20.2 reveals the failure rate of the multichip module interconnection failures only; Table 20.3 does the same for the single-chip-per-package memory. It is assumed that each bipolar circuit has 16 leads or beams and each MOS memory circuit has 24 leads or beams. The results indicate that the multichip module has one-sixth the failure rate or six times the mean-time-between-failure.

TABLE 20.1 Failure Rates for Interconnections

Type of Connection	Failure Rate
Ultrasonic aluminum wire bond	0.000030% per 1000 hr
Ultrasonic beam-lead bond	0.000010% per 1000 hr
Beam lead to IC die metalization	0.000001% per 1000 hr
Kovar lead to PCB	0.000010% per 1000 hr

TABLE 20.2 Failure Rate Calculation for Multichip Module

Beam leads to IC die metalization

480 beams × 0.000001 = 0.00048% per 1000 hr

Ultrasonic beam–lead bond

480 beams × 0.00001 = 0.0048% per 1000 hr

Kovar leads to PCB

26 leads × 0.00001 = 0.00026% per 1000 hr

0.00554% per 1000 hr

TABLE 20.3 Failure Rate Calculation for Single-Die-per-Package System

Ultrasonic aluminum wire bond	
960 bonds × 0.00003 =	0.02880% per 1000 hr
Kovar leads to PCB	
480 leads × 0.00001 =	0.00480% per 1000 hr
	0.03360% per 1000 hr

There are three reasons for the increased reliability exhibited by the multichip module. First, the addition of aluminum beam leads to the integrated-circuit die substitutes a highly reliable interconnection (beam-to-die) for the ultrasonic wire bond. Second, the aluminum-beam bond to the aluminum metalization on the substrate is a more reliable connection than the wire bond to the package. Whereas the beam bond is more reliable because of its greater mass and its rectangular shape, the wire bond is flattened considerably during bonding to ensure sufficient contact area. This deformation of the wire results in weak points in the wire from reduced cross-sectional area and is a source of failure for the packaged die. The beam, however, has more than twice the mass of the wire and, being rectangular, does not require disfiguration to ensure a reliable bond. Third, the added level of integration afforded by the module reduces the number of package leads from 480 to 26; increased relibility is thus obtained because there are fewer interconnections.

Still other aspects of beam-lead assembly, apart from reliability figures on interconnections, result in more reliable operation. A significant difference between beam-lead and standard wire-and-chip assembly is the elimination of the die-attach operation. In the single-die-per-package system, die attachment is performed by placing the IC chip on a glass or metal preform, heating the package to the eutectic temperature of the preform, and scrubbing the die into the now molten die-attach material. This operation is a source of rejection for parts because some dies are not properly attached or are broken or chipped. Failure modes such as cracks in the die that propagate only after various types of mechanical and temperature stressing are not detected until many hours of operation have passed. In addition, the high temperature to which the IC chip is subjected can cause electrical parameter drift.

A beam-lead assembly attaches the die to the substrate through the ultrasonic bonding of the beam to the substrate. The energy to form the bond is applied directly to the beam, and because the aluminum beam is ductile, very little energy is transmitted to the die. The bond is performed at room temperature, eliminating high-temperature exposure of the die.

Both these characteristics are also advantages of beam leads over flip-chip, solder-bump assembly. The flip-chip approach requires that the chip be elevated to the eutectic temperature of the solder bumps, whereupon a scrubbing action or application of pressure is administered to secure the bond. Another advantage of beam leads is that they can be visually inspected for bond integrity, since they are external to the periphery of the chip.

Index